What is Food?

This volume brings together contributions that provide a snapshot of current food research. *What is Food?* acknowledges the many dimensions of food, including its social, cultural, symbolic and sensual qualities, while also being material in that it is fundamental to our survival.

The collection addresses contemporary challenges and reflects the concerns of funders and researchers working in the broad field of the sociology of food: dietary health, sustainability, food safety and food poverty. Reflecting broader academic trends, the chapters are moreover concerned with interdisciplinarity, the analysis of change, data reuse and the use of social media as data. The book includes empirical evidence from around the UK, Denmark, Sweden, Switzerland and Taiwan and addresses food both as a lens through which to examine these wider social relationships, processes and social change and as a primary subject.

The contributions will be of interest to a wide range of students and researchers looking for a cutting-edge insight into how to frame and study food in areas related to the sociology of food, health, risk, poverty, sustainability and research methods.

Ulla Gustafsson is Principal Lecturer in Sociology in Department of Sociology, University of Roehampton. She has published work on school meals in the UK, and her research interests include young people's food practices.

Rebecca O'Connell is Reader in the Sociology of Food and Families, Thomas Coram Research Unit, UCL Institute of Education. She is co-author, with Julia Brannen, of *Food, Families, and Work* (2016) and Principal Investigator of the European Research Council-funded study "Families and Food in Hard Times".

Alizon Draper is Reader in Food and Public Health at the University of Westminster. She has a background in both social anthropology and human nutrition and has been involved in extensive research in the UK and internationally on many topics relating to food and nutrition.

Andrea Tonner is Senior Lecturer in the Department of Marketing at the University of Strathclyde and Co-Convenor of the BSA Food Studies Group. Her research considers food marketplaces and how consumers negotiate food consumption for health and well-being.

Sociological Futures

Series Editors:
Eileen Green ... *Professor Emerita, Teesside University*
John Horne ... *Waseda University, Japan*
Caroline Oliver ... *Associate Professor of Sociology, UCL Institute of Education*
Louise Ryan ... *Professor of Sociology, University of Sheffield, Vice Chair of the BSA*

Sociological Futures aims to be a flagship for new and innovative theories, methods and approaches to sociological issues and debates, and 'the social' in the 21st century. This series of monographs and edited collections was inspired by vibrant wealth of British Sociological Association symposia on a wide variety of sociological themes. Edited by a team of experienced sociological researchers, and supported by the BSA, it covers a wide range of topics related to sociology and sociological research and features contemporary work that is theoretically and methodologically innovative, has local or global reach, as well as work that engages or reengages with classic debates in sociology bringing new perspectives to important and relevant topics.

The BSA is the professional association for sociologists and sociological research in the United Kingdom, with its extensive network of members, study groups and forums, and its dynamic programme of events. The Association engages with topics ranging from auto/biography to youth, climate change to violence against women, alcohol to sport, and Bourdieu to Weber. This book series represents the fruits of sociological enquiry, reaching a global audience, and offering a publication outlet for sociologists at all career and publishing stages, from the well-established to emerging sociologists, BSA or non-BSA members, from all parts of the world.

For more information about this series, please visit: www.routledge.com/ Sociological-Futures/book-series/SOCFUT

What is Food?

Researching a Topic with
Many Meanings

Edited by Ulla Gustafsson,
Rebecca O'Connell, Alizon Draper
and Andrea Tonner

Routledge
Taylor & Francis Group

LONDON AND NEW YORK

First published 2020
by Routledge
2 Park Square, Milton Park, Abingdon, Oxon OX14 4RN

and by Routledge
605 Third Avenue, New York, NY 10017

First issued in paperback 2020

Routledge is an imprint of the Taylor & Francis Group, an informa business

British Library Cataloguing-in-Publication Data
A catalogue record for this book is available from the British Library

Library of Congress Cataloging-in-Publication Data
Names: Gustafsson, Ulla (Ulla S. B.), editor.
Title: What is food? : researching a topic with many meanings / edited by Ulla Gustafsson, Rebecca O'Connell, Alizon Draper, Andrea Tonner.
Description: Abingdon, Oxon ; New York, NY : Routledge, 2020. | Series: Sociological futures | Includes bibliographical references and index.
Identifiers: LCCN 2019031978 (print) | LCCN 2019031979 (ebook) | ISBN 9781138387690 (hardback) | ISBN 9780429426100 (ebook)
Subjects: LCSH: Food—Social aspects. | Food habits. | Nutritional anthropology. | Food supply.
Classification: LCC GN407 .W43 2020 (print) | LCC GN407 (ebook) | DDC 394.1/2—dc23
LC record available at https://lccn.loc.gov/2019031978
LC ebook record available at https://lccn.loc.gov/2019031979

ISBN 13: 978-0-367-72666-9 (pbk)
ISBN 13: 978-1-138-38769-0 (hbk)

Typeset in Goudy
by Apex CoVantage, LLC

Contents

Tables

Figures

Images

Contributors

Editor biographies

Ulla Gustafsson is Principal Lecturer in Sociology in Department of Sociology, University of Roehampton. She has published work on school meals in the UK, and her research interests include young people's food practices.

Rebecca O'Connell is Reader in the Sociology of Food and Families, Thomas Coram Research Unit, UCL Institute of Education. She is co-author, with Julia Brannen, of *Food, Families, and Work* (2016) and Principal Investigator of the European Research Council-funded study "Families and Food in Hard Times".

Alizon Draper is Reader in Food and Public Health at the University of Westminster. She has a background in both social anthropology and human nutrition and has been involved in extensive research in the UK and internationally on many topics relating to food and nutrition.

Andrea Tonner is Senior Lecturer in the Department of Marketing at the University of Strathclyde and Co-Convenor of the BSA Food Studies Group. Her research considers food marketplaces and how consumers negotiate food consumption for health and well-being.

Author biographies

Julia Brannen is Professor of the Sociology of the Family, Thomas Coram Research Unit, UCL Institute of Education and Fellow of the Academy of Social Science. Her international reputation ranges over methodology, families, work–family life and intergenerational relationships. Recent books include *Fathers and Sons* (2015) and *Social Research in the Making* (forthcoming).

Punita Chowbey is Research Fellow at Sheffield Hallam University. Her research is concerned with gender inequalities, race/ethnicity, household economies, economic abuse, food and families and parenting in the UK and South Asia.

Alexi Ernstoff is a sustainability consultant at Quantis in Lausanne, Switzerland. She focuses on food systems sustainability and the interface between health and the environment.

Isabel Fletcher is a research fellow at the University of Edinburgh who uses approaches from STS, medical sociology and history of medicine to analyse the use of nutrition research evidence to advise governments, focusing on the areas of obesity, food security and sustainable diets.

Val Gill is a mixed methods researcher and has worked in health research for over ten years. She has written chapters for reports including the National Diet and Nutrition Survey and the Scottish Health Survey.

Laurence Godin is a postdoctoral researcher in sociology at the University of Geneva. Her work focuses on food, sustainable consumption and everyday life. She is also interested in health, the body and social norms.

Katy Gordon is a doctoral researcher at the University of Strathclyde exploring the community food sector, with a focus on alternative approaches to emergency food aid and use of the social enterprise model. Katy has a background in public health nutrition with research experience working in the third sector.

Judith Green is Professor of Sociology of Health in the Social Sciences & Urban Public Health Institute, Department of Population Health Sciences, King's College London. She has researched and published widely on health research methods, public understanding of risk and sociology of and for public health.

Ming-Tse Hung is a PhD candidate in the Department of Sociology at the University of Edinburgh. His PhD dissertation looks at the discourse of food education in Taiwan, explaining how it problematises real food as its object and develops a different way of knowing food.

Abigail Knight is Lecturer in Sociology at the Thomas Coram Research Unit, UCL Institute of Education. As well as teaching, she is a qualitative researcher with children and families and has published previously about the lives of young people in care, disabled children and families living in poverty.

Hayley Lepps is currently a senior research executive at Ipsos MORI. She predominantly works on the GP Patient Survey, a survey sent out to over a million people each year on behalf of NHS England, to find out how patients feel about their GP practice. Hayley was previously a researcher at NatCen.

Lydia Martens is Professor of Sociology at Keele University. She has a longstanding interest in consumption, domestic practices and family life. Her most recent book, *Childhood and Markets: Infants, Parents and the Business of Child Caring*, was published in 2018 (Palgrave Macmillan). She has also published in *Families, Relationships and Societies*, *The British Journal of the Sociology* of Education, *The Journal of Consumer Culture, Sociology, Sociological Research Online* and the *International Journal of Social Research Methodology*.

Ylva Mattsson Sydner is Professor at the Department of Food Studies, Nutrition and Dietetics at Uppsala University. Her main area of research is foodservice in the public sector. She has especially focused on vulnerable groups such as old people and disabled people.

Jessica Paddock is Lecturer in Sociology at the University of Bristol. Her research explores the interaction of everyday life practices, food consumption and social differentiation in the context of environmental change. Her most recent work has been published in *Poetics, Sociology, Journal of Peasant Studies* and *Journal of Rural Studies*.

Christine Persson Osowski is Associate Senior Lecturer at the Department of Food Studies, Nutrition and Dietetics at Uppsala University. Her main area of research is the Swedish school meal. Other research interests include children and sustainable meals as well as food and religion.

Caireen Roberts is a registered nutritionist (public health) and has worked in food and diet research for over 15 years. She has written chapters for reports including British Social Attitudes and Health Survey for England.

Marlyne Sahakian is Assistant Professor of Sociology at the University of Geneva. Her interest is in the sociology of consumption related to sustainability issues, including energy usage, food consumption and collective well-being. She is a founding member of SCORAI Europe, the sustainable consumption research and action network.

Eleanor Shaw (MA, PhD) is Professor of Entrepreneurship at Strathclyde Business School. She has published several articles on social enterprise and is an advocate of the social benefits of entrepreneurial mindsets.

Alan Warde is Professor of Sociology in the School of Social Sciences, University of Manchester, and a professorial fellow of Manchester's Sustainable Consumption Institute (SCI). Research interests include the sociology of consumption, the sociology of culture and the sociology of food and eating. His most recent book is *Consumption: A Sociological Analysis* (Palgrave, 2017).

Jennifer Whillans is Research Fellow at the University of Bristol. Her expertise lies with time-use research, and she has an interest in the (de)synchronisation and coordination of people and practices that occur in daily life. Her current project is funded by the British Academy: (De)synchronisation of People and Practices in Working Households: The Relationship between the Temporal Organisation of Employment and Eating in the UK.

Juliette Wilson is Senior Lecturer in the Department of Marketing at the University of Strathclyde. Her research interests are focused around alternative and non-mainstream markets and the practices engaged in by all stakeholders in these markets.

Acknowledgements

The BSA Food Study Group collaborated with the SOAS Food Studies Centre in the preparation for the 2017 conference that gave rise to this collection. The editors wish to thank Harry West, the then chair of the SOAS Food Studies Centre, who drafted the call for papers, as well as the other members of the conference committee – Annie Connolly, Elizabeth Hull (co-opted member) and Hannah Lambie-Mumford – for making it a success. We also wish to thank all the contributors to the conference who provided such a rich discussion of this increasingly prominent area of research. Our appreciation extends also to the BSA and in particular Sandria Charalambous for organising the event and to the University of Westminster for hosting it. We are pleased to present the work in the Sociological Futures Series and thank the series editors for valuable comments.

Introduction

*Ulla Gustafsson, Rebecca O'Connell,
Alizon Draper and Andrea Tonner*

Food research in the context of contemporary challenges

Food has been described by the sociologist Anthony Winson (1993, p. 1) as 'the intimate commodity'. It is inherently material: food is fundamental to survival and getting fed is in many ways a practical matter (Harris, 1986). Yet food is also a social and cultural good. It is deeply symbolic (Levi Strauss, 1970) and an important way in which social distinctions and hierarchies are drawn (Fischler, 1988). Food is also sensual: it is not only good to think with but good to eat (Abbots, 2017). Tastes may be cultural but preferences are also personal (Nettleton and Uprichard, 2011). The multidimensionality of food and its embeddedness in everyday life therefore make food a productive lens for examining social realities. But they can also make studying food and eating – let alone bringing about changes in patterns of food production and consumption – complicated. This book takes as its central focus an exploration of the diverse meanings of food and how we interrogate it. It is based on a rich range of empirical contributions that employ different framings of, and means of researching, food. In particular, this volume offers current analysis of topical areas and emergent interests in sociology.

Food is a relatively recent addition to the scope of sociological studies. This is partly to do with the way boundaries between disciplines have been delineated: consumption and its consequences have been the preserve of nutrition while geography and other disciplines have traditionally focused on production (Beardsworth and Keil, 1997). However, with the realisation that food is integral to most areas of sociological interest, it has quickly become a substantially explored topic (Coveney, 2006; Murcott, Jackson, and Belesco, eds., 2016; Warde, 2016; Poulain, 2017). In the *Handbook of Food Research* Murcott, Jackson and Belesco (2016, p. 1) point to the 'burgeoning' field of food research as evidenced by the increase in journals, textbooks and collections now devoted to food related areas. While there are a number of textbooks on the sociology of food (e.g. Beardsworth and Keil, 1997; Germov and Williams, 2008; Guptil, Copelton, and Lucal, 2013) a sociology of food as such is difficult to delineate despite considered discussions identifying several theoretical developments (Poulain, 2017; Warde, 2016). This is partly to do with food relating to a range of areas of, and interests in, sociology. In recent years there are developments of particular theoretical approaches to food and eating, for example, social practice (Warde, 2016) together with a focus on the self (De Solier, 2013) and a sociology of anxiety (Jackson, 2015). Further, the attention on food is noted in sociological works considering its relation to femininity and feminist approaches (Cairns and Johnston, 2015), the body (Abbots and Lavis, 2013; Carolan, 2012), health (Wills, Draper, and Gustafsson, eds., 2013), power (Martschukat and Simon, 2017) and poverty (Coveney and Caraher, 2016). The range of areas within sociology touched by the study of food is therefore considerable. However, in the main each text within the sociological literature limits itself to one particular field of interest.

Drawn from the British Sociological Association (BSA) Food Study Group[1] conference in London, UK, in 2017, the chapters in this collection address contemporary challenges and reflect the concerns of funders and researchers working in the broad field of the sociology of food at this time: dietary health, sustainability, food safety and food poverty. Reflecting broader academic trends that are also related to the nature of the challenges and contemporary funding environment, the chapters are moreover concerned with interdisciplinarity, the analysis of change, data reuse and the use of social media as data. The chapters include empirical evidence from around the UK, Denmark, Sweden, Switzerland and Taiwan. The book addresses food both as a lens through which to examine these wider social relationships, processes and social change and as a primary subject. Together, the chapters offer fresh insights into the methodological affordances and complexities of studying food, engage with 21st-century concerns revealing current shifts in key sociological topics and provide theoretical insights in addressing important challenges.

Studying food: changing theories, concepts and methods

As Warren Belasco notes in the foreword to Peter Jackson and colleagues' *Food Words*, 'food is so relevant to so many key categories of inquiry and experience that we can't even begin to cover the subject' (Belasco, 2015, p. ix). Given this breadth and diversity, describing the field of food research and how it has changed over time is challenging and the subject of a number of overviews (e.g. Mennell, Murcott, and van Otterloo, 1992; Belasco, 2008; Poulain, 2017; Warde, 2016). Addressing research on eating (and excluding food production), Alan Warde (2016) notes its centrality within ethnographic research and highlights important contributions from social anthropology including structuralist studies (Levi-Strauss, 1970; Douglas, 1972) as well as the analysis of how food and eating is related to social hierarchies (Bourdieu, 1979). He also notes the important contributions of feminist studies on the division of labour (e.g. De Vault, 1991) to understandings of 'domestic food provisioning', a concept that neatly bridges so-called public and private domains. However, Warde identifies cultural analysis as dominating food research for a long period of time during the latter part of the 20th century. He argues, 'the foci of the study of eating came to be personal and ethnic identity, nation building and national identity, food scares, eating out, consumer movements and migration, all wrapped up more or less in the problematic of globalisation and consumer culture' (2016, p. 28). At the start of the 21st century, developments in material culture studies and practice theories, together with phenomenological approaches, have emerged as important approaches in the study of food and eating (Warde, 2016).

These changing theoretical and conceptual concerns are reflected in methodological strategies. For example, recent examples of empirical research on food and eating that take a 'practice' approach use a range of methods to understand the activity of food and eating and its embeddedness in everyday routines and social relations. For instance, MacDonald, Murphy and Elliott (2018) used a combination of diaries and interviews to explore more meaningful public health interventions in family food practices. Drawing on the method advanced by Zimmerman and Wieder (cited in MacDonald, Murphy, and Elliott, 2018), they asked respondents to keep an audio diary after an initial interview and then conduct a follow-up interview. They found that 'aspirations and priorities' characterised the initial interview while 'constraints and compromises' were found in the diaries (2018, p. 11). Informed by a focus on practices, the method distinguished between the 'performance' found in the diaries and the 'entity' located in interviews. They concluded that public health initiatives need to consider the social context of food practices as well as the emotions and care-giving associated with food. Similarly, O'Connell and Brannen (2016) used a range of methods to examine children's food practices in the context of parental employment, incorporating a qualitative longitudinal approach to examine change over time.

Reflecting wider trends in the social sciences, a growing interest in the senses and sensory methods is reflected in the domain of food and eating. This includes not only a resurgence in interest in food preferences, tastes and the sensuality of food within academia, but also within communities of policy and practice (see, for example, the SAPERE method [e.g. Kähkönen et al., 2018] and the work of the Flavour School in the UK). The anthropologist Abbots (2017) employs an ethnographic approach in exploring the matter and meaning of food together with how it affects our senses. Her ethnography is informed by phenomenological as well as practice approaches and treats both physical and oral performances and their interplay as data.

In addition to the growing importance of phenomenological approaches, the interdisciplinary subfield of 'food pedagogy' (Flowers and Swan, 2015) has gained prominence in recent years. Studies of food as a means of and subject to educative practices include research on school food and learning activities within and beyond the classroom (e.g. Leahy and Wright, 2016). Many of these studies draw on poststructuralist approaches to understand how people and food are discursively constructed as particular kinds of subjects. Cappellini, Harman and Parsons (2018) adopted a 'biopedagogical' approach, based on Foucault's concept of 'biopower' to analyse power relations through the material medium of the school lunchbox. They demonstrate how the lunchbox serves as a biopedagogical object in which multiple levels of surveillance become visible and consider how these influences are reinterpreted and resisted by mothers.

This brief selection of recent studies illustrates some of the range in how the object of study has been conceptualised as well as in the methodologies adopted when studying food. The chapters in this volume build on some of these approaches both conceptually and in their choice of methods. Employing a range of research strategies including secondary analysis, qualitative longitudinal, ethnographic and visual approaches, the studies included in the volume add interesting and varied contributions to a number of these concerns. They address food and eating as a changing social practice, a resource to be managed in a time of austerity, a source of anxiety in an age of 'risk' and the subject of policy intervention in an era of growing health inequalities and concerns about sustainability. Together they demonstrate the multifaceted nature of food and how its study may illuminate multi-layered social realities.

Organisation of the book

The book is organised into two sections, with the first part serving to foreground the contribution these studies make to methodological questions in relation to the study of food and eating. In Chapter 1 Paddock et al. present a rare opportunity in revisiting a previous UK study two decades later and provide much insight into the methodological challenges this presented. This is quite different from a traditional longitudinal study where the methods would be replicated as the authors have updated the theoretical focus of the research while adapting the methods used. They employ a mixed methods approach that also involves re-analysing the original data in order to provide insights into the changes that have taken place to practices around eating out between the two study times. It is a study that frames the practice of eating as the primary focus rather than using food as a lens; eating is understood as a particularly pertinent example of a wider concern with social practices and their trajectories over time.

In Chapter 2 food is viewed multidimensionally as a lens onto, and dimension of, material and social deprivation. O'Connell et al.'s approach to food poverty in the UK seeks to overcome the shortcomings of research into food poverty that limits itself to focus on the household while ignoring the differential experience of its members. By analysing particular family cases, the authors are able to open up the 'black box' of the family to gain insight into the meaning of food poverty for both adults and children and identify the conditions that are important in understanding their experiences.

In Chapter 3 Godin et al. grapple with a contemporary methodological issue, namely that of how to tackle interdisciplinarity in food research where food again is the focus. This group of Swiss researchers has explored multiple perimeters to their work into 'healthy and sustainable diets'. The collaboration between sociology and environmental sciences presented challenges as well as complexities, and the authors identify 'tradeoffs' required for such work. They conclude, in the area of sustainable diets, relying on one discipline is insufficient and interdisciplinary solutions will need to be developed.

The final chapter in the section uses food as a lens to explore political tensions in a changing social climate and acknowledges a more recent source of data, namely social media. In their Swedish study, Persson Osowski and Mattsson Sydner examine social media debates in Sweden on blog and internet forum posts. They identify tensions between a more individualised demand for school food provision in view of religiously sanctioned food and special diets on the one hand and school meals as a collective welfare provision on the other.

In the second section we have gathered a diverse collection of chapters that represent changes and challenges to the framing and understanding of food. Chapter 5 tackles the topical concern of how the public engages with the food system. There has been much emphasis upon public anxiety with the food chain, but Draper and colleagues identify that there are also many who would prefer not to be so informed about food provenance or 'risks'. The option of 'not-knowing' where food comes from has received little attention in research; here the authors identify its relevance for governance of and trust in the food system.

In Chapter 6 Fletcher picks up the topic from the first section of sustainable diets and traces its relatively recent appearance in the policy literature. While nutrition has been associated with healthy eating advice and individual consumption, sustainability relies on engaging with policy areas focused on production. To combine the two, individuals have recently been advised to eat less meat or consume local and seasonal produce. By focusing on the comparison of Scotland and Denmark, Fletcher identifies the complexities beneath such rather crude guidance and raises concern about the narrow focus in understanding environmental degradation.

Notwithstanding the complexity in involving sustainability in eating advice, Chapter 7 by Chowbey demonstrates how healthy eating is far from straightforward based on nutrition alone. Indeed, the inherent assumptions in attempts at influencing food practices among South Asians in the UK overlook many factors shaping these and ignore the important social and cultural meanings of food as well as socially embedded practices.

Hung's work demonstrates an approach to dealing with food safety scandals in Taiwan that centres on educating people's tastes. Chapter 8 points to the distinction made between 'fake' and 'real' food and examines attempts at teaching people to sense what is real. Informed by Foucauldian discourse analysis, Hung reveals how knowing about the content of food through labels is being replaced in contemporary food pedagogies by 'personal sensual experience'.

The final contribution from Tonner and colleagues addresses the perennial concern of health inequalities and discusses attempts to tackle these in Scotland. The two case studies of community food initiatives in Chapter 9 demonstrate some positive impacts with regard to accessing affordable healthy food. However, they also reveal tensions in the organisations' engagement with social enterprise models given their charitable roots. The authors further describe challenges for the initiatives in recording 'impacts' upon health inequalities which, despite the influential model of the social determinants of health, tend to be monitored at the level of the individual.

Together, these contributions provide a snapshot of current concerns and interests in food research. They capture the methodological breadth employed in food research and engage with contemporary conceptual debates in tackling a wide range of issues.

Note

1 The BSA Food Study Group was established to encourage the sociological analysis, both theoretical and empirical, of all aspects of food production and consumption. Further details, including a history of the group, may be found here: www.britsoc.co.uk/groups/study-groups/foodscoff-scottish-colloquium-on-food-and-feeding-study-group/about/

References

Abbots, E.-J. (2017). *The agency of eating: Mediation, food and the body*. London Bloomsbury.

Abbots, E.-J. and Lavis, A. (eds.) (2013). *Why we eat, how we eat: Contemporary encounters between foods and bodies*. Abingdon: Routledge.

Beardsworth, A. and Keil, T. (1997). *Sociology on the menu*. London: Routledge.

Belasco, W. (2008). *Food: The key concepts*. Oxford: Berg.

Bourdieu, P. (1979). *Distinction: A social critique of the judgement of taste*. Cambridge, MA: Harvard University Press. Translated by Richard Nice.

Cairns, K. and Johnston, J. (2015). *Food and femininity*. London: Bloomsbury.

Cappellini, B., Harman, V. and Parsons, E. (2018). Unpacking the lunchbox: Biopedagogies, mothering and social class. *Sociology of Health & Illness*, 40(7), pp. 1200–1214.

Carolan, M.S. (2012). *Embodied food politics*. Abingdon: Routledge.

Coveney, J. (2006). *Food, morals and meaning: The pleasure and anxiety of eating*. 2nd ed. Abingdon and New York: Routledge.

Coveney, J. and Caraher, M. (2016). *Food poverty and insecurity: International food inequalities*. Switzerland: Springer International Publishing.

De Solier, I. (2013). *Food and the Self: Consumption, production and material culture*. London: Bloomsbury.

DeVault, M.L. (1991). *Feeding the family: The social organisation of caring as gendered work*. Chicago: Chicago University Press.

Douglas, M. (1972). Deciphering a meal. *Daedalus*, 101(1), pp. 61–81.

Fischler, C. (1988). Food, self and identity. *Social Science Information*, 27, pp. 275–292.

Flowers, R. and Swan, E. (eds.) (2015). *Food pedagogies*. London: Routledge.

Germov, J. and Williams, L. (eds.) (2008). *A sociology of food & nutrition: The social appetite*. 3rd ed. Oxford: Oxford University Press.

Guptil, A., Copelton, D. and Lucal, B. (2013). *Food & society: Principles and paradoxes*. Cambridge: Polity Press.

Harris, M. (1986). *Good to eat: Riddles of food and culture*. London: Allen and Unwin.

Jackson, P. (2015). *Anxious appetites: Food and consumer culture.* London: Bloomsbury.

Kähkönen, K., Rönkä, A., Hujo, M., et al. (2018). Sensory-based food education in early childhood education and care, willingness to choose and eat fruit and vegetables, and the moderating role of maternal education and food neophobia. *Public Health Nutrition,* 21(13), pp. 2443–2453.

Leahy, D. and Wright, J. (2016). Governing food choices: A critical analysis of school food pedagogies and young people's responses in contemporary times. *Cambridge Journal of Education,* 46(2), pp. 233–246.

Levi-Strauss, C. (1970). *The raw and the cooked.* London: Jonathan Cape.

MacDonald, S., Murphy, S. and Elliott, E. (2018). Controlling food, controlling relationships: Exploring the meanings and dynamics of family food practices through the diary-interview approach. *Sociology of Health and Illness,* 40(5), pp. 779–792.

Martschukat, J. and Simon, B. (eds.) (2017). *Food, power, and agency.* London: Bloomsbury.

Mennell, S., Murcott, A. and van Otterloo, A. (1992). *The sociology of food: Eating, diet and culture.* London: Sage.

Murcott, A., Jackson, P. and Belesco, W. (eds.) (2016). *The handbook of food research.* London: Bloomsbury.

Nettleton, S. and Uprichard, E. (2011). "A slice of life": Food narratives and menus from mass-observers in 1982 and 1945. *Sociological Research Online,* 16(2), p. 5.

O'Connell, R. and Brannen, J. (2016). *Food, families and work.* London: Bloomsbury Academic.

Poulain, J.-P. (2017). *The sociology of food: Eating and the place of food in society.* London: Bloomsbury. Translated by Augusta Dörr.

Warde, A. (2016). *The practice of eating.* Cambridge: Polity Press.

Wills, W., Draper, A. and Gustafsson, U. (eds.) (2013). *Food and public health: Contemporary issues and future directions.* Abingdon: Routledge.

Winson, A. (1993). *The intimate commodity.* Toronto: Garamond Press.

Part I

Studying food

Chapter 1

Revisiting 'Eating Out'

Understanding 20 years of change in the practice in three English cities

Jessica Paddock, Jennifer Whillans,
Alan Warde and Lydia Martens

Introduction

The original Eating Out project was a study of the consumption of food outside the home, based on extensive original research carried out in England in the 1990s (Warde and Martens, 2000). In 2015, we took what is a rare opportunity in the social sciences to revisit the study, returning to the same three cities – London, Preston and Bristol – to explore changes and continuities in such a practice over time. As will be shown in this chapter, adaptation of the methodology is the result of balancing the requirements of internal validity demanded of repeat studies with re-use of research instruments adapted to the contemporary landscape of market provision and social practice.

A focus upon practices marks a theoretical turn in the sociology of consumption, whereby the kinds of expressions of individual identity play announced by the 'cultural turn' make way for accounts that give less prominence to the agency of social actors, proposing instead that performances of a practice, such as eating, rely more upon automated and practical senses of reasonable action than calculation or deliberation. The unthinking ways in which social actors repeat performances and adjust these according to conditions demanded by various situations lead practice-sympathetic accounts to suggest that people are the carriers of practices, rather than conscious deliberative actors. This 'practice turn' (Schatzki, 2001; Warde, 2016) marks a new phase in sociological research on consumption, and indeed, on eating (Warde, 2016). Eating is understood to be a highly complex, weakly regulated and routinised activity performed by social actors which in turn create the structures upon which these performances are reproduced. Such a turn to practice brings greater theoretical sophistication to a long tradition of research that has framed food as a lens through which to view other domains of social life (Douglas, 1966), including commensality (Fischler, 2011), gender relations (DeVault, 1991; Martens, 1997), sociability (Jacobs and Scholliers, 2003; Julier, 2013; Díaz-Méndez and García-Espejo, 2014), social differentiation and taste (Mennell, 1985; Warde, 1997; Johnston and Baumann, 2010; Cappellini, Parsons, and Harman, 2016; Ray, 2016; Paddock, 2016), deprivation and social exclusion (Wills and O'Connell, 2018) and many more. Understood as a practice, eating is brought to bear as an analytic concept in itself. We extend this lens to the practice of dining out: a field relatively understudied in the UK with the exception of Lane (2010, 2018) and Burnett's (2004) social history of the practice.

To explore changes and continuities in the practice of eating out over time, we take inspiration from the technique of what Burawoy (2003) calls the 'focused revisit'. This involves revisiting sites studied at an earlier time but is distinguishable from a re-analysis or the updating of previous studies. The purpose of a revisit is to explore and explain variation in what is observed, without being enslaved by the rules that govern 'replicable' research. By applying the principles of an ethnographic revisit to a mixed method study of 'eating out' and 'eating in', we were able to re-engage with the topics highlighted by the first visit but bring to it fresh theory and literature to deal with the conceptual priorities of today. In this case, we frame eating out in practice theoretical terms and also address debates in sustainable consumption. To do so, we inevitably open up dialogue between the 1995 and 2015 studies, noting interconnections, developments and departures.

As such, this chapter is mainly methodological in nature. However, we hope that you – the reader – will find it a refreshing change from textbook research methodology, which Glucksmann (2000, p. 1) says focuses 'overwhelmingly on problems, ethics and practicalities of collecting material but offer much less guidance on what to do with it'. We act upon this provocation by presenting a more dialectical and much messier phase in the research process, a stage that is rarely exposed to external scrutiny. We expose the points 'in between' data collection and the polished presentation of findings, which often have the appearance of 'fully constituted knowledge'. More specifically, this chapter explores the logic of revisiting Eating Out and reflects upon the prospects and challenges afforded by this exciting opportunity. It elucidates a number of the challenges – over which there was much head-scratching and agonising – in conducting a sociological revisit, an opportunity which is rare, thus making methodological waters even more unchartered.

We proceed by outlining Burawoy's (2003) theory of reflexive ethnography, which we apply to our combinations of interview and survey techniques of data collection. Not presuming reader familiarity with the original project, we begin by outlining the 1995 research design. With this acting as a backdrop, we explain the design of the 2015 revisit, underscoring points of departure, while emphasising challenges encountered in combining two datasets. With all the ingredients measured out, so to speak, we proceed to demonstrate 'in between' stages of research by discussing two concrete examples: (1) the meaning of eating out and (2) ethnicity and ethnic style cuisine. These two examples demonstrate the challenge of synthesising material – both qualitative and quantitative, from 1995 and 2015 – that very often did not neatly fit together. We conclude the chapter by returning to Burawoy's (2003) four principles of reflexivity.

The logic of revisiting

For Burawoy (2003), the focused revisit is epistemologically grounded by principles of reflexive ethnography. This approach, itself inspired by Bourdieusian approaches, seeks to 'disentangle the movements of the external world from the researcher's own shifting involvement with that same world, all the while recognising that the two are not independent' (p. 646). Concerns about realism over constructivism – whether the changes we observe in the social world are the result of forces external to the researcher or are the product of how the observer understands and constructs that world through the theoretical lenses they bring to the field – are balanced by reflexive engagement with the interaction of both inevitabilities. Crucially, these principles of reflexivity are as applicable to mixed methods research design as they are to ethnography.

A 'revisit' involves returning to sites studied at an earlier time but is distinguishable from a re-analysis or the updating of previous studies. That is, rather than controlling the conditions of the research to the extent that we are permitted only to develop debates in line with the theoretical orientation of the prior study, the research design is flexible in so far as the researchers may look backwards to the past as well as at the present with fresh theoretical lenses, should they wish to do so. This has indeed been common practice in anthropology, where ethnographers return to the sites of classic studies – studied by themselves or by others – conducting empirical fieldwork anew and systematically comparing findings to those of their disciplinary ancestors. They even return to their own sites after sufficient time has elapsed, perhaps because events of social, economic, political or cultural significance render a revisit necessary or important.

Revisits, Burawoy (2003) argues, vary in purpose along a continuum of constructivist and realist motivations. The first, of a constructivist form, he calls the 'refutational' – where the researcher aims to challenge the claims made by the prior study. The second is concerned with furthering understandings generated by their predecessor, thus 'reconstructing' elements of the prior study. Revisits of a realist kind are incentivised by 'empiricist' and/or 'structuralist' aims. Whiffs of constructivist concerns may indeed permeate revisits of a realist kind, but Burawoy reminds us that these tend to focus upon external forces at work in transforming societies. In this way, studies guided by 'empiricist' goals may consider the changes and continuities evident between two time periods, while 'structuralist' revisits may concern themselves with major events, from wars and political upheaval to famines, extreme weather and natural disasters. Such propensity to revisit our own work, and the work of others, Burawoy argues, is relatively uncommon in sociology. Indeed, we, as sociologists, might learn from anthropologists' readiness to revisit studies central to their own canon.

Doing so here, we do not reject the findings of the 1995 study but aim to bring (some) new theories to bear upon its findings by revisiting the same places once again, and by turning once more to both the raw data as well as interpretations presented from this work in 1995. We note there are realist concerns to be addressed when accounting for continuities and change between the two periods, including the 2008 financial crisis, subsequent austerity and the rise of the casual dining phenomenon within the restaurant industry. Mintel (2015) notes increase in eating out but reduction in spending across the sector, as 'meal deals', 'buy-one-get-one-free' offers and 'early bird' set menus provide opportunities for many whose beleaguered budgets would otherwise see an end to any superfluous spending. Indeed, Deloitte (2011) reported that the informal dining out sector was set to continue growing, an industry category that contained informal waited service establishments; fast food and take-aways; coffee shops and sandwich bars; retail 'grab and go's' such as Spar and M&S Food; pubs; workplace canteens; shopping and leisure centres; and establishments linked to travel such as RoadChef and Upper Crust. Their popularity no doubt bred further market investment in the casual dining restaurant as an ideal type. Foodservice publications point overall to the value for money desired by customers facing cut-backs in spending on leisure activities, but who still wish to find ways to go out and enjoy themselves.

Eating out, albeit in less formal, more inexpensive ways, is understood as one such occasion for 'getting out' on a budget, and the businesses able to capitalise in such a way were forecast to be the most likely to survive the recession following the 2008 financial crisis. While we take seriously the potential impact of the financial crisis upon the practice of eating, the market has been slowly variegating and expanding over a longer period. Buying meals out in restaurants, hotels and cafes has become increasingly common over the last 40 years in Europe and North America (Cheng et al., 2007; Cabiedes-Miragaya, 2017; Díaz-Méndez and García-Espejo, 2017; Holm et al., 2016; Kjaernes, 2001; Levenstein, 1988). Recent studies across Europe and the US tell more about up-market restaurants and their oft-times celebrity chefs (Lane, 2011; Leschziner, 2015; Rao Monin and Durand, 2003), but with limited information about customers. We know rather a lot about what is cooked and sold in restaurants and cafes across the globe (Berris and Sutton, 2007; Jacobs and Scholliers, 2003; Ma et al., 2006), there being a special interest in the significance of the spread of commercial enterprises purveying different national, ethnic and regional cuisines and their connection with processes of migration (Berris and Sutton, 2007; Panayi, 2008; Ray, 2016). There is a minor interest in food connoisseurs in Canada (Johnston and Baumann, 2010) and a somewhat dated literature on the more basic experience of eating out in Europe and the US (Finkelstein, 1991; Wood, 1995; Warde and Martens, 2000; Warde, 2016). We understand that the recursive dynamics through which social practices are (re)produced are evident in the interaction of market structures and consumer behaviours. Indeed, separating changing practices from changing external environments is an interpretive issue we are still grappling with, along with internal problems of measurement, external/internal problems of interpretation and shifting meanings of the practice of eating out itself.

In the section that follows, we present the research designs of the 1995 and 2015 studies of Eating Out by way of introducing the data we are working with at present in our 'revisit'.

The preceding 1995 study

Eating Out was first conducted in 1995 (Warde and Martens, 2000). It was one of 16 projects, funded by the Economic and Social Research Council, forming the research programme The Nation's Diet: The Social Science of Food Choice. At the time, research had been concerned with the nutritional rather than the social aspects of eating out and little was known about eating out as entertainment and as means to display taste, status and distinction. Warde and Martens (2000) conducted one of the first social scientific investigations on the nature and experiences of eating out.

The 1995 research design entailed two phases of data collection. In the first, Martens conducted interviews with 33 interviewees from 30 households, in diverse circumstances, living in Preston and the surrounding area during the autumn of 1994. The sampling was modelled on DeVault (1991) and they sought to speak with 'principal food providers': that is, 'anyone, man or woman, who performed a substantial proportion of the feeding work in the household' (2000, p. 228). Reflecting the prevailing gender division in domestic work, 28 women and 5 men were interviewed (3 men were interviewed on their own and 2 were present in a joint interview with their partner). Concentration on Preston, a city in Lancashire in northwest England, was opportunistic but there was no reason to think Preston highly unusual. Interviewees were recruited through various organisations including a leisure centre, a community association, a tennis club, an environmental group, a primary school, a trade union branch and a national DIY chain store.

Semi-structured interviews were conducted *first* because it was thought that in the absence of prior social scientific enquiry it would otherwise be difficult to construct informative questions for a survey instrument. Interviewees were asked questions about aspects of eating at home including descriptions of household routines and distribution of food preparation tasks. Questions about eating out included the interviewees' understanding of the term, frequency and reasons for using various places and information details about recent eating out experiences. Preliminary analysis was undertaken on the semi-structured interview data in order to design a questionnaire for the second phase. Thereafter, the interviews were analysed in considerable detail, focusing on shared understandings which defined eating events and differential orientations and attitudes towards eating out.

In phase two of data collection, 1001 people aged 18 to 65 were surveyed in three cities in England: London, Bristol and Preston. These cities were chosen to offer contrasts of socio-demographic composition and, putatively, cultural ambience. Preston was included as representing a northern free-standing city without any particularly eccentric characteristics. London was selected in anticipation that its unique features, including its diverse market and greater volume of provision, would prompt distinctive consumption behaviour. Central and suburban areas were chosen to illustrate potential differences between areas of the metropolis. Bristol was selected as an example of a southern, non-metropolitan city with some claim to be culturally heterogeneous. Since no three cities would be representative of all others in England, these sites were deemed as satisfactory as any. The survey was undertaken in April 1995 and was administered to a quota sample which matched respondents to the overall population of diverse local sub areas of the cities by age, sex, ethnicity, class and employment status. Despite not being a nationally representative sample, there is no reason to consider the survey biased in any particular way as a basis for an initial portrait of the practice in an English context.

Survey questions sought to ascertain each respondent's frequency of eating out, types of outlet visited, attitudes to eating out, extensive detail about the nature of the most recent meal eaten away from home and rudimentary information about domestic routines. Socio-demographic data was also elicited in order to explore variation by class, income, age, gender, education, place of residence and so forth. Analysis of this cross-sectional survey data provided a snapshot of the practice of eating out in 1995.

Mixing methods

When conducting research using multiple data sources and methods the challenge is 'to maintain the integrity of the single study, compared to inadvertently permitting the study to decompose into two or more parallel studies' (Yin, 2006, p. 41). We encountered this challenge not only when working with and fitting together survey and interview data but also when deciding when and how to synthesise data from 1995 and 2015. Commitment to the logic of Burawoy's revisit provides some safeguard against the study disintegrating into parallel studies. The aim of the revisit is not simply to provide an update to the earlier study – presenting findings from two parallel studies, of 'then' and 'now', but in our revisit, we maintain the integrity of the single study by returning to the original raw data and fully incorporating it into the extant research process. Furthermore, there are several ways in which this project illustrates the mixing of quantitative and qualitative methods. The integration of these two methodologies is the result of putting the interpretations of provisional analysis from qualitative in-depth interviews – interviewees represent a nested sample of the 2015 survey respondents – into conversation with the survey analysis, and vice versa. There were many instances in which speculation from the interview data informed a foray of the survey data, both 2015 and harmonised, which then fed back into further work with the qualitative data; conversely, interesting and even seemingly uninteresting pictures painted using numerical data prompted further engagement with qualitative data for explanation or repudiation. We later illustrate this dialogic process of analysing data collected via such a mixed method research strategy, in relation to questions of taste, ethnic cuisine and demographic change.

In 2015, we remain faithful to the skeleton of the original research design by visiting the same cities and conducting two phases of data collection, but begin instead with a survey, followed by in-depth semi-structured household interviews.

Survey data

In the first phase of data collection, in April 2015, 1001 respondents were surveyed in London, Preston and Bristol. The original rationale for selecting these three cities still resonates (that is, contrasts of socio-demographic composition and putative cultural ambience), despite the significant population and provision changes within these cities since 1995. The survey administered asked almost identical questions as previously, pertaining to five thematic areas: (1) eating at home, social activities, division of labour; (2) eating out and takeaway food; (3) the last main meal eaten (at a public establishment or at somebody else's home); (4) domestic entertaining; (5) socio-demographic questions (e.g. to ascertain class trajectory, social connections and cultural capital). The few modifications made to the survey were to suit market, technological and socio-cultural realities of a social world of eating 20 years on, such as the broadening of types and styles of cuisine available, the advent of the internet, commonplace use of mobile telephone devices and the commensurate rise of social media communication platforms, which in 1995 belonged more to the realm of sci-fi imaginaries than social realities. In this way, the same – or similar – information has been asked to a different sample of individuals each time. A final question was added to the survey in 2015: we asked respondents whether it would be alright to contact them again for a follow-up interview.

In 2015 the survey was again administered to a quota sample to reflect the demographic profile of the cities studied. Census Output Areas (OAs), typically comprising around 150 households, were selected at random from across the relevant Local Authorities in proportion to size. Output Areas were stratified by the proportion of residents in social grade AB, using estimated social grade from the census. Interviewers were given quotas based on age and working status interlocked with sex, to reflect the demographic profile of the OAs. While respondents to the 1995 survey were aged 18–65 years, an additional quota of 100 respondents above the age of 65 were surveyed in 2015 to better capture the ageing population (Leach et al., 2013). When conducting analysis of change between survey years, these additional cases are removed but are included when exploring 2015 data independently, thus securing internal validity of comparison while also fortifying the sample for the purposes of understanding the practice of eating out in 2015.

Survey harmonisation

Given that the aim of the study was to revisit the original study, it was necessary not only to analyse the 2015 survey data – making straightforward comparisons with the survey findings from 1995 – but to also re-analyse the 1995 survey data. Therefore, we combined the two surveys to create a repeated cross-sectional dataset and we analysed the data – by looking at each survey year – separately and combined for an we analysis of patterns of change over time.

While the survey administered in 2015 asked almost identical questions as in 1995, there were a number of differences between the two surveys; thus, creating a single dataset was not a simple task of constructing a variable to indicate survey year and appending responses from 2015 to those of 1995. Some problems of comparability require more sophisticated adjustments than others. From least to most complex, we experienced the following issues: (1) where question wording and response alternatives were identical, it was not always the case that variables were measured and categorised in the same way in both years and (e.g. for the variable indicating 'city', the value 2 did not represent Preston in both years); (2) some questions found in the 1995 survey were *not repeated* in 2015; and (3) new questions were introduced in 2015 that had not been asked in 1995. To deal with the second and third points we, very simply, assigned a value to the coding scheme to indicate 'the question was not asked in the survey' and, thus, data was missing. The most critical of practical steps, requiring the most sophisticated of adjustments for comparability was that (4) often the question wording remained the same across surveys, but in 2015 the response alternatives were amended. Response categories differed because there were *additional* response items, *disaggregated* response alternatives to enable a more detailed response or an altogether *different* set of response alternatives.

In many ways, the task of harmonising data is a meticulous, practical task, but this fourth set of practical adjustments (dealing with additional, disaggregated and altogether different response alternatives) provoked considerable reflexive engagement in what it means to conduct a focused revisit. Before presenting a detailed discussion of this reflection, in our discussion of three concrete examples (in the 'Making sense of change' section), we explain our approach to qualitative data collection and analysis.

The in-depth interviews

At the time of the first study, relatively little was known about the practice of eating out from an academic point of view. Some excellent monographs describe owning and working in restaurants in the US (Fine, 1996; Leidner, 1993) and the UK (Gabriel, 1988). The 30 qualitative interviews – conducted in 1995 only with Prestonians – served as a basis for the mixed methods research design, but also gathered understandings that would consequently frame the survey from the point of view of respondents.

Twenty years later, the primary purpose of conducting in-depth interviews was a little different. In the 2015 revisit, the survey was conducted first

(April 2015) and in-depth interviews followed in January 2016. Thirty-one in-depth follow-up interviews were conducted across all three cities; ten in each city, with an accidental extra interview conducted in Bristol. Interviewees were invited to speak of the kinds of routines that shape both the typical and atypical weeks/weekends. We asked about frequency of the various eating events mentioned and about their dining companions, as well as their tastes and disgusts alike. We ended by inviting interviewees to speak about their experiences of both entertaining others to meals in their own homes and of being a guest in the homes of others. Conducting a large-scale survey and then subsequently conducting in-depth interview with a small number of people who had completed the survey means that the interview sample was nested within survey sample. Before even stepping foot into the in-depth interview, we knew a considerable amount about each interviewee: socio-demographics, class trajectory, their tastes and much more. The explicit request for specific pieces of information using the survey method elicits information that may not have emerged otherwise and certainly would not have emerged in systematic ways across each narrative account. It allowed us to piece together quantitative responses and qualitative accounts, albeit with the proviso that these accounts were collected nine months apart.

Having asked respondents at the end of the survey whether it would be alright to contact the respondent again for follow-up interview, we had a (seemingly) overwhelming positive response from 731 of our 1101 survey respondents (66 percent). From these responses, we were able to select a pool of interviewees to suit the needs of the revisit. Rather than speak to the 'principal food providers' as in 1995, responses to attitudinal and other relevant questions guided the selection of interviewees. For example, given that we are interested in learning about how eating in and out relate to other modes of food provision, we avoided contacting respondents who claim never to eat out or have little to no interest in food or cooking.

Having access to 731 respondents does not, however, promise the responsiveness of those who are selected. Indeed, closely defined criteria – which in our case include a mixture of ages, ethnicities, genders, class trajectory and enthusiasm for food and cooking – reduces this pond to a pool of less than 100. Reducing this pool to a puddle is the commonplace matter of attrition (Edwards and Holland, 2013), as the final number depends upon willingness of potential respondents, wrong telephone numbers, diary conflicts and final 'no-shows'. As one of us arrived at the agreed place and time of interview, ringing once or twice on the front door bell, the interviewee flew from the back door and into a waiting taxi! Furthermore, while attempting to speak with people with a range of socio-demographic profiles for interview, at times the survey respondent's information – used to inform our selection and the knowledge of which we entered the interview setting – had become outdated by the time of interview. Major life-course events – such as relationship break-down, critical illness and childbirth – illuminated conversations about routines shaping their everyday lives, and change therein.

The next section demonstrates the challenge of synthesising material – both qualitative and quantitative, data from 1995 and 2015 – that very often did not neatly fit together. We discuss the processes using three concrete examples: (1) the meaning of eating out, (2) ethnicity; and (3) ethnic style cuisine. In this discussion, we aim to 'open up' the research process of conducting a focused revisit, showing the messy and sometimes agonising in-between stage between data collection and 'fully constituted knowledge' (Glucksmann, 2000).

Making sense of change: frequency, shifting meanings, expanding tastes

Frequency

To measure frequency of eating out, the survey asked in both years about the number of times the respondents think they have eaten out within the last 12 months and asked many more questions to gather detail about the *last time* they ate a main meal away from home. The survey contains two different means to estimate the frequency of eating out among the sample – a retrospective estimate about behaviour in the last year[1] and a report of the last occasion when the respondent ate a main meal away from home.[2] Both questions were asked in both survey years. Given that the question wording and response alternatives were identical in 1995 and 2015, one would be forgiven for assuming that analysing change in frequency of eating out over the 20-year period was a straightforward task.

Common sense assumptions shared with us by many in the run-up to the beginning of the project anticipated dramatic, or if conservative, at least a discernible increase in frequency of eating out over this period. On the contrary, our quantitative analysis suggested remarkable continuity, with the frequency of eating out remaining relatively steady. Uneasy about this contradiction between our expectation and empirical findings, we entered into a dialogic process that entailed returning to the interpretation of 1995 qualitative data, looking afresh at 2015 interview data, taking insight from each to reinterpret our quantitative findings and inspire further quantitative analysis with the harmonised dataset. This led us to the conclusion that the meaning of the practice has somewhat evolved since 1995, which we discuss in greater detail in the following section, entitled '"Ordinary" meals out'.

To gauge whether features of an event that we would call the 'main meal of the day' are shared by our interviewees, we asked what the term 'main meal' meant to them, generating the following understandings held in common:

> For me it would probably be the meal you have in the evening after work or whatever, that you have with whoever else is in your house really.
>
> – Felicity, Preston

For us it's always a cooked meal as a family in the evening. Sundays might be lunchtime but usually it is a main meal in the evening.

– Nicola, Bristol

So for me a main meal is probably the largest portion size, the most calories, probably hot and I'd probably only have one main meal a day, which for me is probably dinner, unless I know that I'm meeting friends for lunch or I have a special lunch date, otherwise dinner to me would be the main meal.

– Penny, London

Figure 1.1 implies some discrepancy between retrospective recall and more recent experience. If we extrapolate from the reported last meal, the mean frequency of eating out in spring 2015 was probably at least once in 10 days, whereas the retrospective estimate would suggest approximately once every 17 days. Only respondents whose last meal was in a restaurant are included when plotting when the last main meal was eaten out (n = 813); to create a comparable estimate, respondents who never ate out were not included when plotting frequency of eating out (n = 1034). Thus the mean, median and modal response to the retrospective estimate of frequency of eating at a restaurant was 'monthly', whereas last occasion registered 'fortnightly' as mean and median, and 'within the last seven days' being the modal response. Forty-six percent of respondents reported that their last meal out was within the last seven days.

There are several possible reasons why these estimates vary, the most probable being that respondents recall smaller and less significant occasions when asked about their last main meal away from home, whilst their longer-term memory alights on more significant events. Moreover, the in-depth interview offers

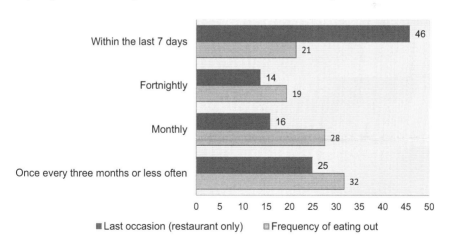

Figure 1.1 Frequency of eating out: instance of most recent occasion and annual estimate (percent).

extended opportunities to recall eating events, and thus sheds some light on what we call the shifting meaning and normalisation of eating out. Additionally, there can be external explanations for variation – such as market shifts resulting from the 2008 financial crisis creating new opportunities to eat out, that commensurately adjust the expectations of eating out, as well as consumers finding occasions to suit these opportunities – which are not solely the result of researchers' interpretations. The explanation we give for the continuity of frequency of eating out is found in conversation between the survey and interview data, where the type of occasions reported and the meanings attributed to these suggest slow evolution in the practice of eating out. Indeed, we are persuaded that, as argued elsewhere by Oriel Sullivan (2004, 1997), that while the metaphors used to epitomise the scale of transformation characteristic of late modern 'runaway' societies – as characterised by Beck, Giddens and Lasch's (1994) dramatic 'juggernauts' and 'volcanoes' of change – fluctuations and variations in everyday practices are small and evolutionary. Such evolution is meaningful, and, we argue, has implications for the performances of practices and the social relations which are produced and reproduced by them. Let us explain a little more.

'Ordinary' meals out

In 1995, Warde and Martens characterised eating out as 'a specific socio-spatial activity, it involves commercial provision, the work involved is done by somebody else, it is a social occasion, it is a special occasion, and it involves eating a meal' (2000, pp. 46–47). Importantly, 'eating out' did not include breakfast or snacks, it was associated with purchase in the commercial sector, and it was, in individual interviewees' words, 'a change from the everyday' and most typically 'a special occasion, dining, in a restaurant or a café, or something' (ibid, p. 45). Events were considered an exception to the quotidian, a source of pleasure and a highly valued opportunity for social interaction.

With this in mind, we sought to characterise the meaning of eating out in 2015, working with the suspicion carried since conducting these interviews in person – that what might characterise a 'special' event, or indeed 'just a social occasion', which are terms used to capture reasons for eating out in the survey instrument – may have somewhat altered. Using NVivo 11, one of us had already carved the data into numerous themes, under which 'reasons for eating out' refined further the range of reasons and purposes described by interviewees. In this way, a picture or vignette of the continuum of reasons for eating out on behalf of each interviewee emerged, leading us to one of our first findings, that of normalisation and simplification (see Paddock, Warde, and Whillans, 2017, for a full account).

To begin with, asking interviewees about the kinds of event that characterise eating out elicited the following kinds of responses:

> *Probably a restaurant, eating out. It doesn't really matter, as long as you're sitting down. I wouldn't count a fast food chain as eating out, that would be more of a*

snack, like you'd get a Subway or a McDonald's. I wouldn't turn 'round to my Mrs and say I'd take her out for a meal and then take her to Subway.

— Tyler, Preston

Not eating in your home. Or a close family member.

— Nadine, Bristol

Dinner. Hot meal at a sit-down restaurant.

— Douglas, London

These responses illustrate the ease with which interviewees were able to define this practice. There is consensus about what counts as eating out, the kinds of foods involved (hot), with whom such events take place, and where. In the survey, respondents were asked whether the reason for their most recent eating out occasion was for (1) A special occasion (SpOc); (2) Just a social occasion (JSO); (3) Convenience/quick meal (C/Q); (4) Business meeting/meal; or (5) Other (specify). Figure 1.2 shows that between 1995 and 2015 the proportion of last meals in restaurants that were described as special occasions has fallen, the proportion described as 'convenience/quick' has increased, while the proportion which are 'just social occasions' and 'business' remains largely unchanged. The shift in restaurant meals has been primarily from special occasions to convenience/quick events.

Looking to the qualitative data, we similarly find an informalisation and simplification narrative across the data and in all three cities. To explore the meaning of more informal and more regular dining, we looked to the range of dining events described by interviewees and distinguished the special from what appeared to be quotidian or what we call 'ordinary' eating out events. While

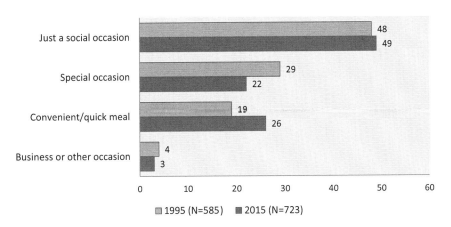

Figure 1.2 Reasons given for eating out on the last occasion at a restaurant (percentages), 1995 and 2015.

special meals should be 'memorable events', most of those described by interviewees are less exceptional. These 'ordinary' meals are shaped and inspired by myriad related practices and are unremarkable and un-exceptional; fewer are planned in advance, they are tied to everyday activities and they are related to everyday responsibilities like 'feeding work' (DeVault, 1991). They are more informal and also more affordable, but nevertheless central to repertoires of eating and sociability.

We further characterise these 'ordinary' events into the 'impromptu' and the 'regularised'. Most interviewees perform a variation of this practice to different degrees, and at different times, according to the social circumstances, and ties to other practices. Releasing oneself from the labour of preparing food as well as other domestic obligations and chores, enters the impromptu event:

> I can't be bothered, so let's go out. It's just easier.
>
> – Nicola, Bristol

Lara (London), who suffers with an ongoing illness, does not always feel well enough to cook, so takes her family out to a 'carvery' restaurant in its place. This, she claims is her way of making 'sure that the family gets together'. Eating out offers an alternative means of accomplishing 'feeding work' (DeVault, 1991; Sullivan, 1997).

Not all 'ordinary' meals out can be characterised as impromptu. Accounts are sprinkled with references to appointments made with others. What we call 'regularised meals' take place at intervals by agreement with kith and kin outside the household, and meals out together are a regular means through which to stay in touch with each other. Often, these are patterns of sociability described as having emerged over time.

> There's four of us usually go out together . . . friends of ours, we usually go out with them once a month or something like that, different restaurants.
>
> – Gerald, Preston

These occasions are particularly important for women:

> By the time you've done everything and everything is ready, by the time you get to it you don't want it.
>
> – Enid, Preston

> We've got lots of exciting things to try, unfortunately it doesn't always appeal to him, so I don't get to try them, but I try them with my Thursday Girls.
>
> – Miranda, Bristol

As we argue elsewhere (Paddock, Warde, and Whillans, 2017), these are opportunities to spend time comfortably with each other without routine domestic chores encroaching on their leisure time (Sullivan, 1997), where their tastes can be indulged without catering to the tastes of their families (Charles and Kerr, 1988) and where the social world of their table cannot be so easily interrupted by anyone other than waiting staff. We suggest, therefore, that over 20 years the meaning of the practice of eating out has evolved. It has become more ordinary as impromptu and regularised social meals provide alternative templates. The normalisation of eating out alters understandings and enables new opportunities for sociability.

Arriving at such interpretations is the result of dialogic analysis of two datasets, in tandem. Triangulating our data sources and analysis in this way is, however, more than a means of simply corroborating results by making the same observations in different settings. Rather, Brannen (2005) suggests four possibilities generated by triangulation that go beyond corroboration. These are named 'elaboration/expansion'; 'initiation'; 'complementarity'; and 'contradiction'. Our revisit elaborates and expands upon both qualitative and quantitative data through gnomic analysis, but as we move through analysis asking different questions of the data, to address different debates pertaining to eating out, we initiate inquiry with one dataset, before complementing initial insights derived from the examination of the first. We are yet to encounter serious conflicts between both datasets, but do find productive tensions between them, for example, in the interpretation of changing meanings against modest changes in frequency observed using the survey instrument.

To illustrate further the messy dialogic process of analysing data collected via a mixed method research strategy, we explain how we came to understand dynamics of taste for various types and styles of cuisine in relation to a shifting landscape of ethnic diversity.

Taste: ethnicity and cuisine type/style

Net migration figures suggest increased ethnic diversity of the British population over the last 20 years (ONS, 2016), seeing a commensurate scholarly interest in understanding the effects of cosmopolitanism (Prieur and Savage, 2013) upon many areas of socio-cultural life in Britain, such as musical taste, dress, food consumption, social integration and so on. Given the renewed affordances of the data in 2015, we join this cause by exploring taste and eating out in 2015, seeking to understand with greater detail the relationship between taste and ethnicity.

In both 1995 and 2015, respondents' ethnicity was identified by asking: 'To which of these (ethnic) groups do you consider you belong?'. In line with official classificatory practices (ONS) in 1995, 9 response alternatives were offered, while in 2015, 19 response alternatives were offered (Table 1.1). Both *additional* response items (e.g. options to identify as multiple/mixed ethnicities) and *disaggregated* response alternatives (e.g. identifying national/regional origin within the identification as 'white') were offered in 2015 compared with 1995.

Table 1.1 Ethnic identification response alternatives in 1995 (left) and 2015 (right)

1995	Freq.	Percent	2015	Freq.	Percent
White	888	88.7	English/Welsh/Scottish/ Northern Irish	765	69.5
Black Caribbean	24	2.4	Irish	12	1.1
Black African	6	0.6	Gypsy or Irish Traveller	0	0.0
Black other	6	0.6	Other white European	59	5.4
Indian	22	2.2	Confirm another white background	24	2.2
Pakistani	14	1.4	White and black Caribbean	16	1.5
Bangladeshi	5	0.5			
Chinese	7	0.7	White and black African	9	0.8
Other	25	2.5	White and Asian	5	0.5
Refused	3	0.3	Confirm any other mixed/multiple ethnicities	8	0.7
Don't know/not answered	1	0.1			
			Indian	60	5.5
Total	1,001	100.0	Pakistani	22	2.0
			Bangladeshi	9	0.8
			Chinese	7	0.6
			Confirm any other Asian background	15	1.4
			African	23	2.1
			Caribbean	17	1.5
			Confirm any other Black/African/ Caribbean	11	1.0
			Arab	14	1.3
			Any other ethnic group	1	0.1
			Confirm any other ethnic group	12	1.1
			Vague or irrelevant	7	0.6
			Refused/not answered	5	0.5
			Total	1,101	100.0

Both the greater detail captured in 2015 and the greater proportion of respondents who identify as a group other than white British unlock the potential to explore eating out and ethnicity. Indeed, this is in part a response to critique that the conflation of ethnic differences into categories of white/non-white addresses issues of taste only ever from the point of view of the white majority (Ray, 2016).

With reference to cuisine style, in 2015 – as in 1995 – we asked: 'During the last 12 months, in which of the restaurant types and places listed on this card have you eaten on the premises?' In 1995, Chinese and Thai were aggregated and offered as a single response alternative. Japanese and French cuisine were not

Table 1.2 Cuisine styles experienced in the last 12 months in 1995 (left) and 2015 (right)

1995	2015
1. Indian	1. Traditional British
2. Chinese/Thai	2. Indian restaurant
3. American style restaurant/diners (e.g. TGI Fridays, San Francisco Exchange, Hard Rock Café)	3. Chinese restaurant
	4. Japanese restaurant
4. Other ethnic restaurant (e.g. French, Greek, Mexican)	5. Thai restaurant
	6. Italian restaurant (e.g. local Italian restaurant, Zizzi etc.)
5. Vegetarian	7. French restaurant
6. Other	8. American style restaurant/diner (e.g. TGI Fridays, Hard Rock Café)
	9. Other ethnic restaurant (e.g. Greek, Mexican, Middle Eastern)
	10. Vegetarian restaurant
	11. Modern British cuisine (e.g. nouvelle cuisine, molecular gastronomy)
	12. Any other kind of cuisine
	13. None of these

offered as response categories in 1995. Traditional British and Modern British were newly offered responses in 2015 (Table 1.2).

The expansion of response alternatives has enabled the substantive exploration of ethnicity and cuisine styles. However, due to limitations of the data in 1995, it is not possible to conduct quantitative analysis using our survey data that expounds changes in taste of non-white British groups over the 20-year period between 1995 and 2015. Instead, we contextualise our findings in 2015 within other research concerned with the changing landscape of taste, the increased diversity of the population and the relationship between them.

As reported in Warde, Whillans and Paddock (2019), the data suggest that a significant proportion of the sample are conversant with a wide range of different culinary styles, a feature which must impact upon the overall tastes and diets of contemporary Britons. As a point of comparison, 48 percent of respondents reported in 1995 that they had not eaten in an 'ethnic' restaurant of any kind in the last 12 months, but by 2015 that proportion had fallen to 22 percent.[3] In 1995, the propensity to visit restaurants defined by their selling of 'ethnic' cuisine was a strong indicator of social position; people with high levels of cultural and economic capital were much more likely during the previous year to have visited several different types of 'ethnic' restaurant than those with less education and working class occupations (Warde and Martens, 2000, p. 81). It was argued that having a broad familiarity with 'ethnic' cuisines was evidence of culinary curiosity and adventurousness and was a mark of distinction. The data thus confirms a meaningful association between taste and social position (see Warde, Whillans, and Paddock, 2019), for which we depend upon expanded categorisations of variables such as ethnicity and class trajectory.

Conclusion

This chapter has explored the opportunities and challenges of revisiting the study Eating Out, conducted by Warde and Martens in 1995. We have emphasised that the triangulation of two datasets has allowed us to initiate analysis from multiple points of entry, to expand upon lines of inquiry, to elaborate, to sense-check hunches and interpretations both expected and surprising – e.g. the relative stability of frequency of eating out reported in the survey. Indeed, the confidence with which we speak about the data comes from its mixed methods approach, and from the assurance we take from not having only revisited claims made from the 1995 data but from the harmonisation of both survey datasets. Such re-analysis triples the work but also increases the strength of claims made.

Nevertheless, the confidence with which we interpret the data must be met with some reflection. Having taken inspiration from Burawoy's (2003) focused revisit, we promised to consider internal and external explanations for the changes and continuities we observe between these two points in time. Burawoy proposes that the changes and continuities noted between earlier accounts and a later revisit, can be credited with differences in '1) the relation of observer to participant, 2) theory brought to the field by the ethnographer, 3) internal processes within the field itself, or 4) forces external to the field site' (2003, p. 645).

Because we researchers eat out regularly, our own experience probably influences our understanding. However, the observations we make, and our subsequent interpretations, are grounded in conversation with practitioners of eating out other than ourselves, and which are observable in the patterning of responses to variables pertaining to eating practices in 1101 cases. Moreover, having focused upon eating events across three cities, we can speak to the relationship between place, modes of provision and practice. We bring theories to bear upon this data that extend, and in Burawoy's terms, reconstruct the 1995 study in order to empirically observe change and continuity over a 20-year period. Thus, our motivations, and the dilemmas of reflexivity we face, pertain to the realist and constructivist kinds. Crucially, Warde and Martens (2000) documented for the first time the relationship between eating at home, eating at restaurants and eating at the homes of others, which contributes to a body of work that frames eating as a practice (Warde, 2017). We continue to elucidate the patterns and meanings shaping eating out as a practice, and not as a series of individual behaviours. We find that one core dynamic at play is that of normalisation of eating out (see Paddock, Warde, and Whillans, 2017).

Turning to dilemmas of a realist kind, we can account for potential variation internal to the practice of eating. The financial crisis of 2008 – recovery from which saw the implementation of the government's austerity programme – brought consequences felt by households whose budgets have been squeezed by the shrinking of incomes in real terms. This may account for the relatively small increase in the frequency of eating out since 1995, while market responses to austerity have been guided by the desire to provide more casual and in many cases more affordable ways to prop up the sector during difficult times.

These principles of reflexivity highlight the ways in which we are alerted to various reasons for change and continuity in the practice of eating that we seek to explain. Indeed, we open up the research process to scrutiny (Glucksmann, 2000) by speaking both to procedural issues in the analysis as well as to logics guiding ongoing interpretations.

Notes

1 Overall how often have you eaten out in a restaurant, pub, café or similar establishment during the last 12 months, excluding times when you were away on holiday (in the UK or abroad)?
2 When did this occasion take place?
3 Warde and Martens (2000, p. 83). The proportion of the sample who had visited an Indian restaurant increased from 33 to 44 percent, for Chinese and/or Thai the increase was from 29 to 42 percent, and for Italian, 31 to 52 percent.

References

Beck, U., Giddens, A. and Lasch, S. (1994). *Reflexive modernization*. Cambridge: Polity Press.

Berris, D. and Sutton, D. (2007). *The restaurants book: Ethnographies of where we eat*. Oxford: Berg.

Brannen, J. (2005). NCRM methods review papers, NCRM/005. *Mixed Methods Research: A Discussion Paper*. Available at: http://eprints.ncrm.ac.uk/89/ [Accessed 1 June 2018].

Burawoy, M. (2003). Revisits: An outline of a theory of reflexive ethnography. *American Sociological Review*, 68, pp. 645–679.

Burnett, J. (2004). *England eats out: 1830 – present*. Harlow: Pearson.

Cabiedes-Miragaya, L. (2017). Analysis of the economic structure of the eating-out sector: The case of Spain. *Appetite*, 119, pp. 64–76.

Cappellini, B., Parsons, E. and Harman, V. (2016). Right taste, wrong place: Local food cultures, (dis)identification and the formation of classed identity. *Sociology*, 50(6), pp. 1089–1105.

Charles, N. and Kerr, M. (1988.) *Women, food and families*. Manchester: Manchester University Press.

Cheng, S.-L., Olsen, W., Southerton, D., et al. (2007). The changing practice of eating: Evidence from UK time diaries, 1975 and 2000. *British Journal of Sociology*, 58(1), pp. 39–61.

Deloitte/BDRC Continental (2011). *The taste of the nation: The future trends for the going out market*. No place of publication identified.

DeVault, M. (1991). *Feeding the family: The social organization of caring as gendered work*. Chicago: University of Chicago Press.

Díaz-Méndez, C. and García-Espejo, I. (2017). Eating out in Spain: Motivations, sociability and consumer contexts. *Appetite*, 119, pp. 14–22.

Douglas, M. (1966). *Purity and danger: An analysis of the concept of pollution and taboo.* London: Routledge.

Edwards, R. and Holland, J. (2013). *What is qualitative interviewing?* London: Bloomsbury.

Fine, G.A. (1996). *Kitchens: The culture of restaurant work.* Berkeley: University of California Press.

Finkelstein, J. (1991). *Dining out: An observation of modern manners.* New York: New York University Press.

Fischler, C. (2011). Commensality, society and culture. *Social Science Information,* 50(3–4), pp. 528–548.

Gabriel, Y. (1988). *Working lives in catering.* London: Routledge.

Glucksmann, M. (2000). *Cottons and casuals: The gendered organisation of labour in time and space.* London: Routledge.

Holm, L., Lauridsen, D., Bøker Lund, T., et al. (2016). Changes in the social context and conduct of eating in four Nordic countries between 1997 and 2012. *Appetite,* 103, pp. 358–368.

Jacobs, M. and Scholliers, P. (2003). *Eating out in Europe: Picnics, gourmet dining, and snacks since the late eighteenth century.* Oxford: Berg.

Johnston, J. and Baumann, S. (2010). *Foodies: Democracy and distinction in the gourmet foodscape.* New York: Routledge.

Julier, A.P. (2013). *Eating together: Food, friendship, and inequality.* Chicago: University of Illinois Press.

Kjaernes, U. (2001). *Eating patterns: A day in the lives of Nordic peoples.* Lysaker: National Institute for Consumer Research.

Lane, C. (2010). The Michelin-starred restaurant sector as a cultural industry: A cross-national comparison of restaurants in the UK and Germany. *Food, Culture and Society,* 13(4), pp. 493–519.

Lane, C. (2011). Culinary culture and globalization: An analysis of British and German Michelin-starred restaurants. *British Journal of Sociology,* 62(4), pp. 696–717.

Lane, C. (2018). *From taverns to gastropubs: Food, drink, and sociality in England.* Oxford: Oxford University Press.

Leach, R., Phillipson, C., Biggs, S., et al. (2013). Baby boomers, consumption and social change: The bridging generation? *International Review of Sociology,* 23(1), pp. 104–122.

Leidner, R. (1993). *Fast food, fast talk: Service work and the rationalisation of everyday life.* Berkeley: University of California Press.

Leschziner, V. (2015). *At the chef's table: Culinary creativity in elite restaurants.* Stanford: Stanford University Press.

Levenstein, H. (1988). *Revolution at the table: The transformation of the American diet.* New York: Oxford University Press.

Ma, H., Huang, J., Fuller, F., et al. (2006). Getting rich and eating out: Consumption of food away from home in urban China. *Canadian Journal of Agricultural Economics,* 54(1), pp. 101–119.

Martens, L. (1997). Gender and the eating out experience. *British Food Journal,* 99(1), pp. 20–26.

Mennell, S. (1985). *All manners of food: Eating and taste in England and France from the middle ages to the present.* Oxford: Blackwell.

Mintel (2015). *Eating out review UK: June 2015.* Available at: http://store.mintel.com/eating-out-review-uk-june-2015?cookie_test=true [Accessed 8 Dec. 2016].

Paddock, J. (2016). Positioning food cultures: Alternative food as distinctive consumer practice. *Sociology*, 50(6), pp. 1039–1055.

Paddock, J., Warde, A. and Whillans, J. (2017). The changing meaning of eating out in three English Cities 1995–2015. *Appetite*, 119, pp. 5–13.

Panayi, P. (2008). *Spicing up Britain: The multicultural history of British food*. London: Reaktion Books.

Prieur, A. and Savage, M. (2013). Emerging forms of cultural capital. *European Societies*, 15(2), pp. 246–267.

Rao, H., Monin, P. and Durand, R. (2003). Institutional change in Toque Ville: Nouvelle cuisine as an identity movement in French gastronomy. *American Journal of Sociology*, 108(4), pp. 795–843.

Ray, K. (2016). *The ethnic restaurateur*. London: Bloomsbury.

Schatzki, T. (2001). Introduction: Practice theory. In: T. Schatzki, K. Knorr Cetina and E. von Savigny, eds., *The practice turn in contemporary theory*. London: Routledge, pp. 1–14.

Sullivan, O. (1997). Time waits for no (wo)man: An investigation of the gendered experience of domestic time. *Sociology*, 31(2), pp. 221–239.

Sullivan, O. (2004). Changing gender practices within the household: A theoretical perspective. *Gender & Society*, 18(2), pp. 207–222.

Warde, A. (1997). *Consumption, food and taste: Culinary antinomies and commodity culture*. London: Sage.

Warde, A. (2016). *The practice of eating*. Cambridge: Polity Press.

Warde, A. and Martens, L. (2000). *Eating out: Social differentiation, consumption and pleasure*. Cambridge: Cambridge University Press.

Warde, A., Whillans, J. and Paddock, J. (2019). The allure of variety: Eating out in three English cities, 2015. *Poetics*, 72, pp. 17–31.

Wills, W.J. and O'Connell, R. (2018). Children's and young people's food practices in contexts of poverty and inequality. *Children & Society*, 32(3), pp. 169–173.

Wood, R.C. (1995). *The sociology of the meal*. Edinburgh: Edinburgh University Press.

Yin, R. (2006). Mixed methods research: Are the methods genuinely integrated or merely parallel. *Research in the Schools*, 13(1), pp. 41–47.

Chapter 2

Food poverty in context

Parental sacrifice and children's experiences in low-income families in the UK

Rebecca O'Connell, Abigail Knight and Julia Brannen

Introduction

Food poverty has become a major moral and social concern in Britain (Dowler and O'Connor, 2012). It has been defined as 'the inability to acquire or consume an adequate quality or sufficient quantity of food in socially acceptable ways, or the uncertainty that one will be able to do so' (Dowler, Turner, and Dobson, 2001, p. 2) However, families' and children's experiences of food poverty are largely absent from public discourse (Knight et al., 2018a), and little research takes into account the experiences of children and young people. This is exacerbated by the failure of many large-scale studies of food insecurity to treat the household as a differentiated unit whilst qualitative studies often focus, albeit for good reasons, on mothers. The little research that includes children has often abstracted them from their families. This chapter analyses the experiences of children who live in low-income households where financial and food resources are stretched, as well as those of their parents.

The first part of the chapter describes the context in which families with children in the UK are experiencing poverty. The second section describes the conceptual and methodological approaches of the study, on which this chapter draws, and those who took part in the study. The next section analyses food shortage at the level of the household in terms of 'parental sacrifice' and child hunger before describing four cases in depth: two in which the mother's sacrifice protected children from the direct experience of food poverty and two in which it did not. The final section briefly discusses the affordances and limitations of the methodological strategy adopted and the implications of the findings for policies that seek to address food poverty.

The UK context

The effects of economic recession and welfare retrenchment in some European countries following the 2008 global financial crisis have led to a significant decline in household incomes and, increasingly, vulnerability to poverty

(Matsaganis and Levanti, 2014). This is especially true in the UK where, under the guise of reducing public debt, so-called austerity measures from 2010 have hit many British people hard. The Welfare Reform Act passed in 2012 has introduced progressively harsher cuts to welfare spending such as the freezing of Child Benefit and the introduction of a 'benefit cap' on the overall value of benefits a family can receive, including a limit to the amount of housing benefit that can be claimed, despite rising rents. Under this act, the introduction of the under-occupancy penalty (known as the 'bedroom tax'), the implementation of Universal Credit and further deductions in the access to Employment and Support Allowance for disabled people have produced much controversy, as well as hardship. In April 2017, the two-child limit on the child element of Child Tax Credit and Universal Credit came into effect. This policy discriminates against large families and severs the link between need and support that is the foundation of a 'just and compassionate welfare state' (Sefton and Tucker, 2018, p. 1). This situation has resulted in rising levels of relative poverty, particularly among families with children (Mack, 2017). Some suffer more than others. Since 2010, lone parents in the lowest 20 percent of the income distribution have lost 8 percent of real income; couples with children, about 11 percent (Marsh et al., 2017, p. 41).

This income loss is reflected in rising levels of food poverty and demand for emergency food aid among families. The provision of food parcels to households with children by charitable food banks has grown considerably since the introduction of welfare reform, austerity measures and fluctuating living costs (Lambie-Mumford and Green, 2015). Whilst food banks have become a symbol for food poverty in 'Breadline Britain', those who manage to get by without recourse to food banks, perhaps by turning to friends and family or who live in ever-increasing debt, by and large receive little attention (Knight et al., 2018a).

Food poverty, like income poverty, is not evenly distributed. According to the Households Below Average Income figures for 2013/2014 (DWP, 2015), families particularly at risk of income poverty in Britain include lone parent families, workless households (plus those with only one adult in paid work) and those with three or more children. Research based on Joseph Rowntree Foundation's Minimum Income Standard (MIS) found that large families (households with three or more children) and lone parent families are most likely to struggle to meet the budget standard for a socially acceptable, healthy diet and be at risk of food poverty (O'Connell et al., 2019). Parents and children in lone parent households make up the largest proportion of those helped by Trussell Trust food banks, and children living in large families are also over-represented (Loopstra and Lalor, 2017).

It has long been found that some parents (particularly mothers) deny themselves food if there is not enough to go round to protect their children and partners from going without (e.g. Pember Reeves, 1913; Spring Rice, 1939; Dowler and Calvert, 1995). As the social anthropologist Pat Caplan (1996, p. 218) notes, the expectation that mothers sacrifice their food intake for others is one that many 'women have internalized to the point where it becomes second nature, and they may even articulate a preference for less valued food'. Food poverty also involves social exclusion: parents are unable to offer hospitality or allow children to have friends home for a meal (Ridge, 2002).

Research also suggests that children in some low-income households also skip meals or report going to bed hungry (e.g. Hall et al., 2013) and that some children, particularly those who receive free school meals, experience 'holiday hunger' (Machin, 2016; Lambie-Mumford and Sims, 2018), an issue that has recently been taken up by local authorities, charities, politicians and the media (Forsey, 2017). However, research on food poverty among children in the UK is limited, small-scale and has generally been carried out by non-government organisations (e.g. Harvey, 2014). We know little about how children and young people living in different types of families experience and think about food poverty or how children and their families negotiate food and eating in conditions of disadvantage and inequality.

Concepts, methodology and methods

Reflecting feminist and sociological work from the 1980s (e.g. Brannen and Wilson, 1987; Wallman, 1984), in the study on which this chapter draws, the household is viewed as a resource system. Here we examine intra-household distribution – how resources are shared and managed between different family members – including between parents and children. As sociologists of food, families and childhood, we view children as agents in their own lives and at the same time acknowledge that, as minors, they are financially dependent on, and legally the responsibility of, their parents. Whilst children have some control over what they eat, their consumption of food and other goods cannot be understood as independent of their families (Cook, 2008).

In the study, we understand families, food and eating as practices that are embedded in and reproduced through everyday routines and social relations (O'Connell and Brannen, 2016). This approach has had implications for the study's methodology, resulting in the deployment of a range of qualitative methods. We have adopted a 'multiple perspectives' approach (Ribbens McCarthy, Holland, and Gillies, 2003; Harden et al., 2010), in which data from parents and children are brought together, providing a rich and complex picture of the everyday practices constituting family life and the place of food within it (O'Connell, 2013).

The study

The study, titled Families and Food in Hard Times, is funded by the European Research Council. It examines the extent and experience of food poverty among children and families in three European countries, including Portugal and Norway as well as the UK. Alongside secondary analysis of large-scale data, in all three countries qualitative interviews were carried out with young people, aged

11–15 years,[1] and their parents or carers. Forty-five families took part in both the UK and Portugal and 43 families in Norway. The study was carried out in two contrasting study areas in each country in the years 2015–2017. In the UK, 30 of the sample lived in an inner London borough and 15 in an English coastal town in the South East, both areas with high child poverty rates of over 40 percent (after housing costs) (End Child Poverty, 2016). The two UK study areas differ in terms of employment opportunities, housing costs and demographics, with the inner London borough being more ethnically mixed.

After ethical approval was obtained from the funder and the research institution, families in both areas were recruited via a short self-completion survey sent to parents of children in three schools. This was followed by other methods of recruitment, such as referrals from local charities that were providing a variety of resources and help to families and persons in need. Careful attention was given to ensuring ethical standards were adhered to throughout the study (Knight et al., 2018b).

The study aimed to gain insight into these families in as much depth as possible, within the limits of our resources. Semi-structured interviews were conducted with parents (usually mothers, but included a lone father, lone grandmother and male partners in four of the couple families) and young people (30 boys and 21 girls). The interviews covered the following topics: income and outgoings, food budgets and practices, social lives regarding food and eating, sources of support and perspectives on managing and feeding the family on a low income. Whilst we sought to interview parents and children separately, housing conditions meant this was not always possible. Interviews with parents lasted roughly 1.5 hours on average and with young people around 45 minutes. A subsample of families (12/45) participated in two follow-up visits that involved visual methods, to learn more about foods bought and eaten at home and elsewhere and how any changes in circumstances impacted on food and eating over time (O'Connell, 2013). These visual methods included a tour and photography of the family's kitchen and food storage areas with the parent, followed by an interview with the young person about the photos they had taken of food they had eaten recently in different settings, such as home, friends' homes and in their neighbourhoods. In most cases, and with parents' and young people's permission, interviews were recorded and transcribed verbatim.

As we demonstrate in the chapter, the analysis of the information gathered involved treating the families as 'whole cases' – viewing all the material from one family together and setting it in social context. Our aim was to take account of the intersecting characteristics of each household, for example, the type of household, the ethnic backgrounds of family members, parents' employment status and family size. Based on extensive field notes, we also took account of their immediate neighbourhoods, the settings in which the interviews were conducted and the encounters between the interviewees and interviewer. Through integrating the data generated without subordinating one view of the world to another (Mason, 2006), a sense of the complexity of family life was preserved. The approach to analysing and comparing the households was both rigorous and systematic. Together the research team compared the cases, analysing the commonalities across and the differences between them (Gomm, Hammersley, and Foster, 2000). The study speaks not only to how they managed resources within the intimate spaces of family life but also to the changes taking place beyond the household.

The families

The sample reflects the diversity of ethnic backgrounds. In both study areas, around half the mothers were British, but in the coastal area all the British mothers were white. In the inner London borough there was more diversity among the British mothers (18/30) that included white British and black British as well as British Asian mothers, whilst a third (10/30) of mothers had migrated from outside the EU (West and North Africa). In the total sample, parents in 7 of the 45 families were migrants from mainland Europe and the majority of these (5/7) lived in the coastal area. In the former group, there were several cases in which the parent and children had been cut off from state benefits while the families were in the (lengthy) legal process of applying for 'leave to remain'.

Two-thirds of the sample (30/45) were lone parent families. In 25 of the 45 households one or more parents was in paid work and 16 were reliant on benefits. Four families were not in employment and did not receive benefits, due to their legal status. Half (15/30) the lone parents were in paid work although, as many studies show, paid employment and poverty are not necessarily stable conditions. People moved in and out of employment, benefit regimes and entitlements changed and people's lives were beset by other kinds of life events. Those not captured in large-scale survey research were children in families whose legal status in Britain was problematic (they lacked 'indefinite leave to remain'). The four families in the study without leave to remain were the most deprived: unable to work, they had no recourse to public funds (NRPF) at the time we met them and were totally reliant on charity. Although not typical of those living in poverty in the UK, their experiences reflect the devastating effects of successive governments' regulation of welfare benefits as a tool for controlling immigration (O'Connell and Brannen, 2019).

Comparing the incomes of our families to the pattern of household income nationally,[2] most were in the bottom two quintiles: 21/45 were in the lowest and 20/45 were in the second lowest. A few families (4/45) were in the middle-income group. Thirty-nine families were in debt, which included credit cards, loans, rent or utility arrears and high interest rent-to-own purchases.

Food poverty within households: children's and parents experiences

In defining poverty we draw on the 'relative deprivation' approach developed by Peter Townsend in the 1950s and 1960s in Britain (Townsend, 1979) as encompassing multiple dimensions – social exclusion as well as material deprivation

Table 2.1 Parents going without food by child reports of going hungry (N = 45)

		Child feels hungry		
		YES	NO	
Parent goes without	YES	**(A) Parent and child go without and are hungry** (12 cases)	**(B) Parental sacrifice means child gets enough to eat** (11 cases)	23
	NO	**(C) Parent gets enough to eat but child does not** (1 case)	**(D) Neither parent nor child goes without enough food or is hungry** (21 cases)	22
		13	32	45

(O'Connell, Knight, and Brannen, 2019 and O'Connell et al., 2019). In this chapter, we focus on the 'quantity' dimension of food poverty for the families, in particular how far they go without enough decent food, how they 'get by' (Lister, 2004) and the extent to which children as well as parents are directly affected by food shortage. Table 2.1 shows our analysis, in which we divided the families into four groups:

A   Parent and child both go without enough food and are hungry; some of these families are literally 'destitute', by any definition of the term.
B   Parent goes without and is hungry but child gets enough to eat.
C   Parent gets enough food but child does not (a pattern related to the fluctuating nature of [food] poverty and the different times in which the parent and child were interviewed).
D   Neither parent nor child goes without enough food or is hungry; an outcome of mothers investing time in cooking and making significant compromises in the nutritional quality of food.

In Group A both children and parents say they go without enough to eat. In Group B are families in which parents manage to protect their children from hunger through sacrificing their own food. The largest group (D) is of families in which parents and children say that they have enough food in terms of quantity. However, this does mean that everyone is well fed. Rather they 'get by' – that is use everyday strategies and made compromises in making sure everyone is fed, including buying and eating food of poorer nutritional quality. Group C contains an outlier explained by the dynamic nature of poverty and the methodology.[3] For reasons of space we focus on pattern A and B only in this chapter and give two cases for each.

(A) Parent and child go without enough food and are hungry

In around a quarter of the families, children went hungry despite the parent (mother) sacrificing her own food intake to try to protect her children. All but one family in this group were headed by lone mothers. Nearly all these families were in the lowest income quintile, reflecting the fact that, in most cases, they were not in formal paid employment. Four of the seven mothers were migrants who were neither allowed to work nor to claim benefits because their legal status meant they had no recourse to public funds. These four families were therefore dependent on charity and were in absolute poverty. We now look at contrasting experiences of children from two families with different living conditions, both from inner London.

Charlie

Charlie, aged 15, is a white British boy who lives with his mother, a lone parent, in the inner London borough. Charlie's mother lost her teaching job two years ago and is on Job Seekers Allowance that, together with Child Tax Credit and Child Benefit, amounts to £166 a week. The family lives in social housing and the rent is covered by Housing Benefit. They are in the lowest income group. Charlie's mother spends about £30 per week on food. She has resorted to using the local food bank twice recently because of a large gas bill and bank overdraft charges. She never eats breakfast and regularly skips meals. Charlie's mother has also cut down on the quality of food since she became unemployed. She cannot afford to buy fresh fish or cuts of meat that she used to be able to buy. She says she is *'skipping some really nice things that I used to really like'*. Her son, she says, eats a lot and, typical of teenagers, is always hungry. She is very concerned because her son is growing fast, but also because he eats a lot of processed food. As a mother she feels she is letting him down.

Charlie himself says he is always hungry. He does not eat breakfast at home and at school he uses his allocation of £2 for free school meals at break-time. By lunchtime he is hungry again. However, he is restricted to a small sandwich costing £1.80 because the cost of the larger baguette he would prefer exceeds his free school meal allowance and, in any case, he has usually spent his money by then. So he goes home at lunchtime to look for food, typically instant noodles or pasta with pesto or butter, *'but sometimes there's no food'* (Image 2.1).

After school, he hunts for food at home again, *'I just see what's in the house, look in the fridge, look in the cupboard. Look in the fridge again, hoping more food's just appeared'*. At supper time his mother says she pads out meals with pulses, but this is difficult because Charlie doesn't like lentils and chick peas, so he won't eat them. His mother ends up making two meals. She cooks him sausages, pizza, pasta or noodles. He says, *'I just eat like pretty much the same thing [every day]'*. After dinner, Charlie feels hungry again, and late at night, after his mother has gone to bed, he eats more pasta or noodles, or a frozen burger. The foods he puts together often make for unconventional meals (Image 2.2). Food is more plentiful, Charlie says, when his mother gets her benefits and does the shopping at Iceland.[4]

Image 2.1 Charlie's fridge. Charlie's mum said, '*I like to have a full fridge, cupboards ... it's depressing having an empty cupboard. It's horrible to open the fridge ... and Charlie will open the fridge and go "What's in here?" like "Oh okay". I'm not really providing enough.*'

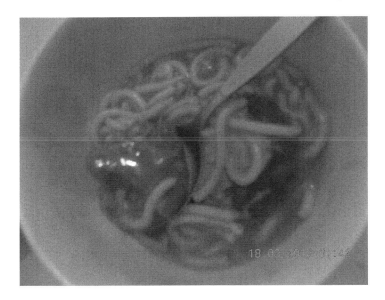

Image 2.2 An unconventional meal made by Charlie. '*There was only one burger left and spaghetti, so I just mixed it together*'.

Image 2.3 Pasta with salt. Asked whether Charlie ever had it with anything on it, oil or butter, he said he did but there was neither: '*On this day there was like nothing left in the house apart from pasta*'.

In the questionnaire, Charlie said he 'often' went to both school and to bed hungry. When there is insufficient food at home to fill him up he fantasises about a '*big meal*' whilst eating '*little bowls of pasta*' saying, '*even if I do get food it's not like a sufficient amount, so I'd still be hungry after I eat it. Especially as I've got a fast metabolism as well, I just like to eat*' (Photo 2.3). Asked when gets hungry, he answered, '*I'm hungry right now*'. Unsurprisingly, Charlie aspires to become a chef: '*I just kind of like the idea of it, and I like the fact that you can just eat whenever you wanted in a restaurant*'. Children's imagined futures are shaped in different ways by their current circumstances (Attree, 2006); for Charlie, becoming a chef is a way of 'getting out' (Lister, 2004).

Amara

The second family lives in a hostel in inner London. Amara, also aged 15 years, was born in southern Europe and lives with her mother, who is originally from North Africa. They recently moved to the UK in order, as her mother said, '*to give my daughter education'*. After initially living with friends, they were placed in temporary accommodation in one room in a very large hostel, but when we interviewed them, they were facing eviction. Amara's mother is registered at the Job Centre but has not been able to find suitable employment. The family receives no benefits. Instead they rely on friends while sometimes the mother does informal work in return for '*cash in hand*'.

There is no food budget. Amara and her mother live hand to mouth. The mother went to a food bank but on the last occasion was turned down because she had used up her quota of three visits a year. She protested unsuccessfully: '*I said sorry well we have to eat. Well we're [not] eating just three times a year. I'm sorry to say that, I'm sorry. Well we're eating every day, humans. She said, "this is how it works"'*.

Both Amara and her mother act in ways that suggest a great deal of sacrifice and altruism. Each considers the needs of the other and this extends to the limited quantity and quality of food. As Amara's mother says, '*I say "well okay I can struggle, I can starve for my daughter", you understand, I want her to have proper education, proper stuff'*. The mother goes without food during the day and reserves what little there is for the evening when they eat together:

> *Sometimes like I don't [eat] nothing just – I wait for my daughter to come at home and we have sandwich which we have, well tin of tuna or something like that, you understand? I can starve all day long waiting for her, like then we can share what we have at home. This is how it is, you understand? . . . Morning I had coffee and that's it really, yeah, soft drinks or something or some toast. This is my day.*

Amara says she often goes to both school and bed hungry. Like her mother, she sacrifices her own food intake:

> *I skip meals to share with my mum (inaudible) . . . for example, I skip my meal to wait for her to come back and at least we can have the same amount of food. . . [we] starve together through the whole day, so at least we will have had something to eat.*

When her mother has no money, she gives any she has to her mother: '*I try to keep (inaudible) put money together and just help each other, that's what we do. . . . Instead of (inaudible) keep the money, come home and give them to my mum, or just share them'*.

Amara is not eligible for free school meals and until recently did not receive them. So she ate nothing at school: '*I used to starve in school because . . . well I couldn't manage to make sandwiches at home or take crisps or whatever (inaudible) so I was just starving in school for the whole day'*. This affected her school work: '*When I'm hungry I just can't concentrate, it's really, really hard for me to do that . . . so I just need to make my mind up and know that I will eat after 5 hours, 7 hours when I get home'*. Amara's mother eventually spoke to the school, which now provides a free school meal. The allowance (£2 daily), however, does not provide sufficient food to satisfy Amara's appetite, *'a small sandwich is like £1.60'*. Amara would like to take cheaper food from home but this is not an option, *'but when I don't have food at home what am I going to do?'* Although school meals can be vital for children living in poverty, they are not always enough to fill them up.

Amara and her mother both say they enjoy and take a pride in cooking, having lived in a southern European country. They cook Mediterranean dishes and are *'learning to cook English'*. They prefer fresh, home-cooked food. However, as many low-income mothers comment, it is more expensive to buy the ingredients for cooking from scratch than to buy ready-made meals. They eat much less meat and fish than they would like. Moreover, the reality of their life leaves them little scope for preparing homemade food. In their one room they have poor cooking facilities and unhygienic conditions in which to store food (the building has cockroaches).

(B) Parental sacrifice means child gets enough to eat

In other families, parents manage to protect their children from going hungry by going without enough to eat themselves. However, children may be affected indirectly by food shortages.

Around half the parents who sacrificed their food intake were successful in protecting their children from going without food. Just over half of these were headed by lone parents (five mothers, one father and one grandmother) and the rest were couples. In couple households, some mothers said they ate less than their partners, but two fathers in this group who were interviewed also reported eating less or differently from their children. In one family in which the mother said they ate less meat because of a lack of money, the father, who was reluctant to admit they were struggling, said this was a matter of preference.

All but one of these families were in the lowest two income groups. In half of the families, at least one parent was in paid employment, whilst in the other half, the parents were not in paid work because of caring responsibilities, disability or ill health. In half of the cases in which one parent was in paid employment, income fluctuated because they were on variable or 'zero hours' contracts. All but one of the families not in paid employment were in receipt of benefits.

Parental sacrifice to protect children from going without food is in some cases a response to fluctuating income or outgoings. Maddy's grandmother, for example, says she skips food once or twice a week during her granddaughter's school holidays (when there is no free school meal). In the case of Owen (which follows later), an unexpected expense in combination with a variable income means that there is less money for food at times. In other cases, where there is chronic low income, skipping meals and eating less is an established way that parents manage. Faith's father, a lone parent in low-paid full-time employment, skips food in the morning to ensure his four children can eat, saying they are *'the priority, so the children may have something to eat I can starve myself'*. Femi's mother doesn't eat at work, saving what money and food there is so her children don't go hungry at home.

Some mothers – and, as noted, fathers – internalise a preference for eating less or less valued foods; for example, Maya's mother, who says she is '*not bothered*' about going without fish or mince (meat) so the children can eat them. Some mothers also seek to protect children from knowing that they are sacrificing their own food intake. Shaniya's mother says, '*Last week I didn't even eat for four days. . . . And . . . I have to lie to my kids and tell them I've eaten so that they're okay, because as long as my kids are eating then I'm okay*'.

Children in some families, however, worry and feel guilty about their parents' altruism; for example, Bryony, a 13-year-old girl who lives in the coastal area in a home in which food is often scarce (Photo 2.4), says that:

> If there isn't enough food we'll get it and sometimes mum will go hungry and starve and stuff. Even if it's not that much food for me and [brother], it's enough that we've actually had something, whereas mum hasn't, and it gets a bit to the point where we'll start feeling guilty because mum hasn't had anything and we've had it.

Image 2.4 Bryony's mother gets paid (her benefits) tomorrow so this is '*the toughest*' time. However, since M shops daily there is usually '*not a lot*' more food than this. '*There'll be a couple of tins of beans, tomatoes, basic things like that's there as a basis and then I've got all my herbs and spices and stuff, and then I can just go out, buy meat and make something little*'.

Whilst parents' sacrifice may protect children from the direct effects of food shortage then, the indirect effects of poverty penetrate deeply into the 'emotional heartland' of children's personal and family lives (Ridge, 2011). We now analyse two cases in this group in more detail. Both are families living in the coastal area and have two parents, but their circumstances are otherwise quite different.

Emma

Emma, aged 12, is a white British girl, one of four children in a two-parent family in the coastal area. Neither parent is in paid work and the family is in the lowest income group. Their weekly income is made up from benefits – Job Seeker's Allowance (JSA), Child Tax Credits and Child Benefit, as well as additional Housing Benefit to cover the rent. Although the two parents are both in poor health, they have been turned down for Employment Support Allowance (ESA) and Personal Independence Payment (PiP) (both benefits for which those with poor health or disabilities can apply to supplement income). Living in an isolated area with little transport, expensive bus fares and no free or subsidised travel for children, the family prioritises running a car that, along with debt repayments, squeezes their budget.

The family spends about £90–£100 per week on food for the family of six. The children all receive free school meals, but the parents subsidise this for the older two children by giving them about £9 or £10 each every week because the food at school 'isn't enough'. They rely on grandparents' support to cover essential costs (such as cooking dinner for the children) and emergencies (such as the cost of car repairs). The parents also depend on food the maternal grandparents grow in their allotment (Photo 2.5).

Image 2.5 Marrow from Emma's grandparents' allotment. 'We get a lot of fresh veg in season'.

As Emma's parents say:

Father: 'We get a lot of fresh veg [from the allotment] in season. So we've got marrow out there, fresh ones, we get carrots when it's time, we get all sorts of stuff . . . parsnips'.

Mother: 'But if it weren't for that . . . it's so dear now. We could end up living off mince sometimes can't we?'

The family's diet is of a mixed quality. Some meals are based on fresh ingredients, such as stuffed marrow and spaghetti bolognaise; others are cheap frozen foods, like pies and chips (Photo 2.6).

When money is particularly tight, the quality and quantity of food suffer, as Emma's father says, '*You know when you get a bigger bill come in, and that has to be paid – things like the electric and the others, then we find we're cutting down on the amount we buy, but then we're also cutting down on the quality*'. Both parents also skip meals at times or miss out what they regard as 'protein' so that there is enough for the children. The mother says, '*we'll get the chips . . . but we won't get the sausage or the fish*', while the father adds, '*Yeah, they [the children] come first, and it's their money that we're getting it out of basically*'.

Image 2.6 'Spag bol'.[5] Emma's mother said, '*A couple of times a week when the money gets tight we will end up with mince because . . . the amount of stuff we add to it, it makes it enough for two meals*'.

In this household, the children confirm that food shortages do not affect them directly. Emma says she never goes to school or bed hungry because of lack of food at home. She is in fact receiving treatment from a dietitian for being underweight, but her father says this is as a result of her having very little interest in eating. Neither the parents nor the child connect this to the shortage of money and food at home. It is known however that children, particularly girls and young women, can internalise financial stress in different ways (McNeish et al., 2015).

Generally food doesn't interest Emma; she has to be reminded to eat and admits to being a 'fussy' eater. She doesn't always eat breakfast but, when she does, she has a cereal bar and yoghurt drink that the dietitian recommended. She has free school meals. At break time she says she has '*a sausage roll or something*' which costs 80p from the £2 that her parents give her daily as a top up. With the rest of the money '*sometimes I get some more food at lunch or I get a drink at lunch*'. She doesn't think her diet is particularly healthy but questions the messages given to her: '*The teachers that do it don't even believe in five a day*'. Her favourite food is Kentucky Fried Chicken, which she has '*once a month if I'm lucky*'.

Owen

The second case is of Owen, aged 12, a white British boy who lives with his parents and brother, also in the coastal area. Owen has a mild learning disability and a bowel problem that requires regular medication, both of which affect what he eats. Since both his parents are in (low-paid) employment, and there are only two children, the family is less squeezed financially than the previous case. However, Owen's mother's earnings are very volatile and, because they earn above the threshold for tax credits, there is no benefit safety net. His father is employed full time in a supermarket warehouse and his mother works part time, but on a zero-hours contract, as a home carer. Owen's mother's take-home pay varies between £800 and £1100 a month, whilst she says her partner's is a '*solid £1200*'. Owen's mother has to travel long distances in her job, but her work mileage allowance is insufficient and she partly meets the costs of petrol from the £130 per month the family gets in Child Benefit.

Owen's mother says they spend around £125 per week on food and that keeping cost down is the key priority, '*Um, we do struggle. We end up having to go for like the cheap options of the food and stuff like that*'. She says she buys food that is filling rather than nutritious and that '*if cost was no bother, I would, the kitchen would be completely full of fruit and veg. It really would*'.

Having an income that is low *and* unpredictable is stressful and affects food and eating. Owen's mother says, '*I keep on checking my emails because we get our wage slip through our emails*'. When money is very tight, she says they purchase cheaper products, '*on some days, well some months, when we can't afford it we'll go and get the pound for four [burgers]. Yeah, the really cheap, tacky things that are virtually full of fat. Yeah, we do things like that*'. They also eat food that goes further, '*It's usually spag bol, because that stretches. You can get a couple of days out of that*'. Towards the end of the month when there is little money for food, as the mother says, '*We usually run down the freezer*'. However, '*if it's looking rather really, really bare, then as long as the boys are fed then that's how it goes*'.

The parents often go without proper meals whilst the children's food needs are prioritised. At the end of the month and when there are additional costs – for example birthdays, Christmas, the start of term (school uniform) – or unexpected expenses, such as a vet bill, the parents skip lunch and resort to sandwiches or toast in the evenings for themselves, '*You know the reason we're on toast at the moment like I say is because of having the dog put down. That came straight out of his wages. We didn't even do, we couldn't even do a shop*'. The mother says, '*I go past the hunger*' and dismisses this as something she and the children's father are content to do:

> So we cut back. As long as, like I say as long as the kids are fed, we don't care about us. We'll sit, we're happy to just sit there and have toast every evening, so we do cut back a lot.

Owen is not directly affected by shortages of food at home and eats three meals a day, plus snacks. He confirms that he never goes to school or bed hungry because there is not enough food at home. He only likes a narrow range of foods, however, which is a common characteristic of his disability. Asked what he thinks is a healthy diet, he suggests it comprises lots of fruit and vegetables but says his diet is '*not completely healthy*' because he eats '*a lot of bacon*'. Owen's mother and father try to cater to his tastes, saying '*We usually know what he likes. He likes his like jam donuts, we usually buy him a packet of jam donuts a week*'. Because Owen is not entitled to free school meals, his mother gives him £1.50 a day for lunch at school. She is concerned that this is not enough: '*I'm worried because I don't know how big the slice of pizza is or . . . because he's not . . . in my eyes he's not eating a lot. He's a growing lad, he should be eating a lot more than that*'. Whilst Owen's mother says he does not ask for more money, it is known that children moderate and manage their needs to avoid putting stress on their mothers (Ridge, 2002).

Discussion

Having enough to eat of adequate quantity and quality has long been a minimal expectation of what it means to live in a western country. Yet in our study, we found families on low incomes who were unable to feed themselves adequately and others who were barely able to do so. In some families, parents often went without, in order that their children did not go hungry, while in others both parent and child suffered.

In this chapter we have used a case approach to take a close look at the everyday realities of food poverty for children and parents in Britain. We have examined some of the conditions in which children and families are unable to access enough decent food. For some families, poverty is a constant feature of their lives and one that they have long become used to. For others, the reasons why they find it so hard to feed themselves lie in the unpredictability of their (low) incomes, the reasons for which are a direct result of welfare, immigration and employment policies including lack of eligibility for benefits of any kind (no recourse to public funds), loss of job, changes and delays in benefits, lack of permanent employment and variable working hours, ill health and lack of eligibility for disability benefit.

The findings support those of other studies that report parental sacrifice to protect children's food intake. They also demonstrate that children may also sacrifice their own food intake and moderate their needs to act in solidarity with, and support, their parents. Food poverty also creates feelings of worry and shame in children and leads to social exclusion.

Case research that treats the household as resource system highlights the specific conditions in which children and their parents go without and get by. Such an approach is resource intensive; recruiting and interviewing children and parents in families and in-depth case study analysis both require skilled researchers and take time. It demonstrates the importance of *not* abstracting children from their families and the wider determinants of food poverty (Lambie-Mumford and Sims, 2018). For example, whilst interventions that 'feed hungry children', such as breakfast clubs and school holiday schemes, vitally meet immediate needs, they cannot address the causes of food poverty. In addition, like other forms of food aid, they risk further stigmatising and marginalising those who are already materially deprived and socially excluded (O'Connell, Knight, and Brannen, 2019 and O'Connell et al., 2019). The chapter's findings suggest that to address food poverty, government should instead ensure that wages and benefits, in combination, are enough to ensure an adequate standard of living and eating for children and their families. In addition, given that school is compulsory for school age children in the UK, meals should also be universally provided, free at the point of service, to all children in state schools.

Acknowledgement

The research leading to these results has received funding from the European Research Council under the European Union's Seventh Framework Programme (FP7/2007–2013)/ERC grant agreement no. 337977. We are very grateful to the children and parents who generously gave their valuable time to participate in the study and to our colleagues in the UK and international research team.

Notes

1 There were three outliers: one 10-year-old and two who were 16 years old either at first or second interview.
2 Taking into account household size and calculated after housing costs.
3 In this case the mother was interviewed a few weeks before the child. In the intervening period the mother lost her job and they visited a food bank. Whilst the mother did not report lack of food at the first visit, the child reported it at the second.
4 Iceland is a UK supermarket chain selling mainly frozen and low-cost foods.
5 'Spag bol' is colloquial shorthand for spaghetti bolognaise.

References

Attree, P. (2006). The social costs of child poverty: A systematic review of the qualitative evidence. *Children and Society*, 20, pp. 54–66.

Brannen, J. and Wilson, G. (eds.) (1987). *Give and take in families: Studies in resource distribution*. London: Allen and Unwin.

Caplan, P. (1996). Why do people eat what they do? Approaches to food and diet from a social science perspective. *Clinical Child Psychology and Psychiatry*, 1(2), pp. 213–227.

Cook, D. (2008). The missing child in consumption theory. *Journal of Consumer Culture*, 8(2), pp. 219–243.

Department for Work and Pensions (DWP) (2015). *Households below average income*. London: DWP.

Dowler, E. and Calvert, C. (1995). Diets of lone parent families. *Social Policy Research*, 71. New York: Joseph Rowntree Foundation.

Dowler, E.A. and O'Connor, D. (2012). Rights based approaches to addressing food poverty and food insecurity in Ireland and UK. *Social Science and Medicine*, 74(1), pp. 44–51.

Dowler, E., Turner, S. and Dobson, B. (2001). *Poverty bites: Food, health and poor families*. London: Child Poverty Action Group.

End Child Poverty. (2016). Available at: www.endchildpoverty.org.uk

Forsey, A. (2017). *Hungry holidays: A report on hunger amongst children during school holidays*. APPG on Hunger and Food Poverty. Available at: https://feedingbritain.files.wordpress.com/2015/02/hungry-holidays.pdf

Gomm, R., Hammersley, M. and Foster, P. (2000). Case study and generalisation. In: R. Gomm, M. Hammersley, and P. Foster, eds., *Case study method*. London: Sage, pp. 98–115.

Hall, S., Knibbs, S., Medien, K., et al. (2013). *Child hunger in London: Understanding food poverty in the capital*. London: Greater London Authority/Ipsos MORI.

Harden, J., Backett-Milburn, K., Hill, M., et al. (2010). Oh, what a tangled web we weave: Experiences of doing 'multiple perspectives' research in families. *International Journal of Social Research Methodology*, 13(5), pp. 441–452.

Harvey, K. (2014). *Children and parents' experiences of food insecurity in a south London population*. Reading: University of Reading and Kids Company.

Knight, A., Brannen, J., O'Connell, R., et al. (2018a). How do children and their families experience food poverty according to UK newspaper media 2006–15? *Journal of Poverty and Social Justice*, 26(2), pp. 207–223.

Knight, A., O'Connell, R. and Brannen, J. (2018b). Eating with friends, family or not at all: Young people's experiences of food poverty in the UK. *Children & Society*, (32), pp. 244–254.

Lambie-Mumford, H. and Green, M. (2015). Austerity, welfare reform and the rising use of food banks by children in England and Wales. *Area*, 49(3), pp. 273–279.

Lambie-Mumford, H. and Sims, L. (2018). Charitable breakfast clubs and holiday hunger projects in the UK. In: W. Wills and R. O'Connell, eds., *Children & Society*. Special Issue: Children's and teenagers' food practices in contexts of poverty and inequality. *Children & Society*, (32), pp. 244–254.

Lister, R. (2004). *Poverty*. Cambridge: Polity Press.

Loopstra, R. and Lalor, D. (2017). *Financial insecurity, food insecurity, and disability: The profile of people receiving emergency food assistance from the Trussell trust foodbank network in Britain*. London: The Trussell Trust.

Machin, R. (2016). Understanding holiday hunger. *Journal of Poverty and Social Justice*, 24(3): 311–319.

Mack, J. (2017). Chapter 7: Child maltreatment and child mortality. In: V. Cooper and D. Whyte, eds., *The violence of austerity*. London: Pluto Press.

Marsh, A. with Barker, K., Ayrton, C., et al. (2017). *Poverty: The facts*. 6th ed. London: Child Poverty Action Group.

Matsaganis, D. and Levanti, C. (2014). The distributional effect of austerity and the recession in Southern Europe. *Southern European Society and Politics*, 19(3), pp. 393–412.

McNeish, D., Scott, S., Sosenko, F., et al. (2015). *Women and girls facing severe and multiple disadvantage in the UK*. London: LankellyChase Foundation.

O'Connell, R. (2013). The use of visual methods with children in a mixed methods study of family food practices. *International Journal of Social Research Methodology*, 16(1), pp. 31–46.

O'Connell, R. and Brannen, J. (2016). *Food, families and work*. London: Bloomsbury.

O'Connell, R. and Brannen, J. (2019). Food poverty and the families the state has turned its back on: The case of the UK. In: H.P. Gaisbauer, G. Schweiger and C. Sedmak, eds., *Absolute poverty in Europe: Interdisciplinary perspectives on a hidden phenomenon*. Bristol: Policy Press.

O'Connell, R., Brannen, J. and Knight, A. (2018). Child food poverty requires radical long term solutions. *BMJ*, 362, p. k3608.

O'Connell, R., Knight, A. and Brannen, J. (2019). *Living hand to mouth: Children and food in low income families*. London: Child Poverty Action Group.

O'Connell, R., Owen, C., Padley, M., et al. (2019). Which types of family are at risk of food poverty in the UK? A relative deprivation approach. *Social Policy and Society*, 18(1), pp. 1–18.

Pember Reeves, M. (1913). *Round about a pound a week*. London: Virago.

Ribbens McCarthy, J., Holland, J. and Gillies, V. (2003). Multiple perspectives on the 'family lives' of young people: Methodological and theoretical issues in case study research. *International Journal of Social Research Methodology*, 6(1), pp. 1–23.

Ridge, T. (2002). *Childhood poverty and social exclusion: From a child's perspective*. Bristol: The Policy Press.

Ridge, T. (2011). The everyday costs of poverty in childhood: A review of qualitative research exploring the lives and experiences of low-income children in the UK. *Children & Society*, 25(1), pp. 73–84.

Sefton, T. and Tucker, J. (2018). *Unhappy birthday! The two-child limit at one year old*. London: End Child Poverty.

Spring Rice, M. [1939] (1981). *Working-class wives: Their health and conditions*. London: Virago.

Townsend, P. (1979). *Poverty in the United Kingdom: A survey of household resources and standards of living*, Harmondsworth: Penguin.

Wallman, S. (1984). *Eight London households*. London: Tavistock.

Chapter 3

From practices to volumes, from meaning to nutrients

An interdisciplinary approach to healthy and sustainable food consumption

Laurence Godin, Alexi Ernstoff
and Marlyne Sahakian

Introduction

Food systems have significant environmental impacts across supply chains, from production to consumption (Tukker et al., 2006). At the same time, in Switzerland as well as globally, diet is a major contributor to the development of non-communicable diseases such as cardiovascular diseases, diabetes and cancer (GBD, 2015; Risk Factors Collaborators, 2016). As a consequence, there is a growing interest in research and policy action towards a change in food consumption that would address both human health and environmental issues. However, inducing significant changes in consumers' habits has proven difficult, and both research and policy trying to link food consumption, health and sustainability face significant challenges.

While research exists that quantifies various impacts of food systems, there is no one definitive guideline on what constitutes a healthy diet for both people and planet, in terms of human health and environmental protection. Further, research and policies can provide different recommendations, for example regarding the amount of meat to be consumed per day, which are disjointed from people's beliefs, representations and habits. These aspects can be largely affected by cultural context, as highlighted by the Food and Agricultural Organization and its emphasis on integrating cultural preferences into dietary guidelines, in order to be more impactful and aligned with the human right to food (FAO, 2012). Similarly, academic discourse on relevant impact categories – such as biodiversity loss and climate change – often disregard habits and traditions that are culturally embedded, thus finding little traction among the general population.

As part of a research project on 'healthy and sustainable Swiss diets',[1] an interdisciplinary team came together in order to develop new knowledge on the health and environmental impacts of food consumption, building on consumer representations and food consumption practices. Quantifying the impacts of diets on human health and the environment is only one step towards guiding healthier and more sustainable eating habits; our main hypothesis is that it is essential to also understand people's food representations and practices in their everyday lives. Bringing together social practice theory approaches (Dubuisson-Quellier and Plessz, 2013; Sahakian and Wilhite, 2014; Shove, Pantzar, and Watson, 2012), with Life Cycle Assessment (LCA) and health impact assessment (Jolliet et al., 2015; Stylianou et al., 2016), our aim is to better understand what opportunities there are for promoting healthy and sustainable diets in Switzerland.

In the following text, we focus on the methodological challenges and opportunities in integrating social practice theory with life cycle thinking, from theory to methods. First, we discuss the context of our research, by describing three gaps in the study of healthy and sustainable food consumption. We then explore how interdisciplinarity allowed us to make our respective disciplinary approaches more relevant to everyday issues as well as to policy concerns. We finally outline the most important challenges brought upon by interdisciplinarity.

Gaps in the study of healthy and sustainable food consumption

The study of healthy and sustainable food consumption has many facets and wide-ranging ramifications within research and policy. As a consequence, theoretical and applied approaches inevitably face gaps in knowledge that hinder both understandings of, and actions required to improve, food consumption. The research project on which this chapter is based comes as a direct answer to some of these gaps. In the following section, we explore more closely three of them that were instrumental in the development of our research strategy: the disconnect between health and sustainability, the focus on individual food items rather than on diets and the overly individualised understanding of food consumption practices that dominates policy and research.

The disconnect between health and sustainability within policy and research

On the one hand, Swiss policies relating to food *production*, such as farmer subsidies, take some aspects of environmental sustainability (e.g. biodiversity) into account, but are mostly centred on the financial interest of producers and economic growth in a highly competitive market (Federal Statistical Office [FSO], 2015; Organisation for Economic Co-operation and Development [OECD], 2011). Production policies do not obviously relate to nutritional health. On the other hand, food *consumption* guidelines (e.g. food pyramids) present proactive health guidance, which do not explicitly relate back to sustainability concerns regarding production and in theory should not be influenced by financial interests. Interventions (e.g. publicly funded school programmes), however, are often based on a financial rationale in the face of exploding health care systems costs and the high prevalence of lifestyle-induced, noncommunicable diseases (Chastonay et al., 2017; Galani, Schneider, and Ruten, 2007; World Health

Organization [WHO], 2017). Another disconnect between health and sustainability are the policies regarding food labelling and marketing. Whereas health claims and nutritional information are highly regulated, sustainability claims and certifications (which can also be regulated depending on the certification) can be used as sales arguments, in an appeal to people's sense of ethics and moral emotions (Antonetti and Maklan, 2014). As most consumers are ill-informed regarding what makes a product more or less 'environment-friendly' (Tobler, Visschers, and Siegrist, 2011), many fall back on labels without necessarily understanding the criteria for labelling or trusting labels entirely (Godin and Sahakian, 2018).

Given these disconnects across production- and consumption-oriented policies regarding health and sustainability, a growing number of governmental and non-governmental organisations working in the field of health and nutrition are trying to include sustainability in their approach to food and diets. Unable to find definitive answers to their interrogations, they may rely on intuition rather than evidence when formulating recommendations. This often means promoting local and seasonal food consumption as a proxy for sustainable food, even though what is 'local and seasonal' is not always well defined, and may not always result in the best environmental performance and social equity (Born and Purcell, 2006). Given different misconceptions and lack of evidence, there is still a need to answer the key question: what would a healthy and sustainable diet look like?

The project on which this chapter is based is financed by the Swiss National Science Foundation as part of a national research programme on 'Healthy nutrition and sustainable food production', which aims to tackle this problem by generating 'praxis-oriented basic knowledge on how to promote healthy nutrition in Switzerland . . . while minimizing the negative impact on the environment and using resources as efficiently as possible', as stated on their website (SNSF, 2018). In other words, a main objective of the research programme is to produce knowledge on the possibility of combining health and sustainability in food production and consumption policies.

At the international level, the last few years also saw the multiplication of initiatives addressing this challenge. For example, the EAT-Lancet Commission for Food, Planet and Health, which published its final report in February 2019 (Willett et al., 2019), brought together leading academic experts as well as governments and non-governmental organisations. The commission set out to

> scientifically assess whether a global transformation to a food system delivering healthy diets from sustainable food systems to a growing world population is possible, and what implications it might have for attaining the SDGs [Sustainable Development Goals] and the Paris Climate Agreement.
> (Rockström, Stordalen, and Horton, 2016, p. 2365)

Many agencies from the United Nations, such as the Food and Agriculture Organization (FAO) and the World Health Organization (WHO), also support academic initiatives and leading organisations in the field, such as the Food

Climate Research Network (FCRN, March 26, 2018). In this favourable environment, progress is being made in the development of the knowledge necessary to define what a healthy and sustainable diet could look like at different levels (individual, national, international), such as the integration of human health aspects into environmental Life Cycle Assessments (LCAs), as described in the section titled 'Articulating environmental LCA and health impact assessment'.

Moving from individual food items to diets

Next to the need for uniting health and sustainability into a common framework comes the need to shift the focus from food items to diets as a whole. Indeed, a substantial basis of research focuses on 'sustainable foods' or 'healthy foods' in an attempt to answer the question: which *food items* are more environmentally sustainable and healthy (Masset et al., 2014)? However, there is increasing evidence that the more appropriate question to ask might be: which *diets* are more environmentally sustainable and healthy?

Comparing the environmental impacts and the 'healthiness' (e.g. nutritional profile) of individual food items has helped establish fundamental knowledge of food systems – for example, that ruminant meat production is generally more impacting than plant-based foods (e.g. per kilogram), and nutritional content varies greatly across foods (Fern et al., 2015; Heller, Keoleian, and Willett, 2013; Roy et al., 2009). Investigating environmental impacts of individual food items can also be useful when comparing different production practices of the same food item, or across the value chain. Through such work, it is well established that agricultural production is generally the most environmental impactful aspect of food value chains (e.g. in comparison to transport). Comparing the environmental impacts of two different food items, however, leads to limited interpretation with respect to what can be recommended for actual food practices and habits. Due to this limitation, newer work has shifted towards investigating sustainable and healthy *diets*, i.e. the overall daily consumption of an individual or population, by building on such food-specific knowledge (Hallström, Carlsson-Kanyama, and Börjesson, 2015; Heller, Keoleian, and Willett, 2013; Nemecek et al., 2016; Tilman and Clark, 2014; Walker, Gibney, and Hellweg, 2018).

Shifting towards a dietary perspective in sustainability assessments is important for several reasons. First, a dietary perspective considers the quantity of various foods consumed. Instead of comparing two different foods based on impact per kilogram or impact per calorie (regardless of quantity actually consumed in a diet), a dietary perspective puts food into the consumption context. This is relevant because a high-impacting food consumed in low quantities can have a lesser environmental impact than a low-impacting food consumed in high quantities (e.g. grains). Second, a dietary perspective can help understand the overall environmental footprint of an individual's or a population's consumption and guide decision-making towards an overall lesser footprint. Such a dietary perspective should acknowledge that reductions in consumption of one food item – for example, red meat – will likely be compensated by a replacement or substitution by another food item. Without data regarding food substitution preferences, in LCA research substitutions are often addressed as a modelled assumption based on per weight, calorie or volume, which may not reflect actual consumption or be easily interpretable (Ernstoff, Stylianou, and Goldstein, 2017; Eshel et al., 2016; Notarnicola et al., 2017; Tilman and Clark, 2014). Another line of evidence supporting the need for a dietary perspective comes from the health angle. Generally, the 'healthiness' of food or nutrition must be considered in the context of overall consumption. For example, sodium is an essential human nutrient, but overconsumption can lead to an increase in health risks related to cardiovascular disease (GBD, 2015; Risk Factors Collaborators, 2016). From this perspective, foods 'high in sodium' are only unhealthy if they are consumed in a diet that is overall 'high in sodium'.

Given the strengths of a dietary perspective, it is important that future LCA research continues to focus on identifying healthier and more sustainable *diets* – moving away from a focus on food items. More data are needed to understand dietary change for example when a certain food item, such as red meat, is replaced. Finally, understanding dietary change can help indicate how to ensure overall global consumption is within the safe operating space of 'planetary boundaries' – the biophysical constraints of maintaining human life on earth (Campbell et al., 2017). From this perspective, no one food (or diet) can be 'sustainable' unless the entire global system is safely within biophysical constraints.

The pervasiveness of an individualised understanding of food consumption

While health and sustainability should be considered across the whole food system, from the production of foods to their distribution, storage, consumption and final waste, dietary transitions or changes are often individualised (i.e. the claim that an individual chooses their diet). As documented in consumption studies over the past decade, the view of consumption as being motivated by individual decision-making processes has dominated policy arenas thus far (Cohen and Murphy, 2001; Fahy and Rau, 2013). The main assumption is that informed consumers make rational decisions. This viewpoint has been criticised in what has been termed the value-action gap or attitude-behaviour gap: consumers may be aware of what ought to be healthy and sustainable diets, they may even express beliefs or attitudes that are aligned with these perceptions, but this does not always translate into actual practices (Blake, 1999; Kollmuss and Agyeman, 2002; Rau, Davies, and Fahy, 2014; Shove, 2010). For example, a majority of the Swiss population is aware of main dietary recommendations, yet only 30 percent of the population eats the recommended fruit and vegetable intake each day (Federal Office of Public Health [FOPH], 2012). It has also been shown that the Swiss population is not aware of the environmental impacts of food consumption, and that social norms are found to be a stronger influence on personal values and choices (Kamm et al., 2015). Moreover, values and beliefs in one area of consumption (e.g. healthy and sustainable food) may or may not translate into other areas (e.g. socialising, caring for a newborn baby, the world of work).

To go beyond the value-action gap and the idea of rational individual choice, social practice theory approaches have been gaining in popularity among researchers and policy-makers, particularly in relation to (un)sustainable consumption practices – including food. Building on earlier work by Bourdieu (1979), Giddens (1984) and Schatzki (1996), social practices are seen as being made up of three main elements – relating to the material, individual and social dimensions of practices – which come together to form the doings and sayings of everyday life. What makes up these elements of practice can be described in different ways; in one interpretation, the object of study in a practice approach are materials, competences and meanings (Shove and Pantzar, 2005; Shove, Pantzar, and Watson, 2012). Here, food consumption practices – such as preparing a meal or eating – are apprehended as habitual and based on routines. In the context of social practice theory, meanings are about signs and symbols that help reproduce practices. For example, the meaning of the American Thanksgiving meal is structured around a family gathering to give thanks and the cooking and eating of a turkey. If one of these aspects is taken away – the turkey or the family gathering – the meaning of Thanksgiving changes. Thus, meanings are not fixed but dynamic, and are only maintained and reproduced if taken up in practice.

Social practice theories are bringing new perspectives to food consumption studies precisely because they move beyond individual 'rational choices' to an understanding of consumption as meaningful and related to everyday life (Halkier, 2009; Halkier and Jensen, 2011; Jaeger-Erben and Offenberger, 2014; Warde, 2013). The focus on everyday habits and constraints is also emphasised in the Swiss Nutrition Policy report (FOPH, 2012), along with what are termed 'other barriers', such as lack of nutritional knowledge, insufficient information, pricing factors and taste preferences.

While much work has been done on drivers and barriers in relation to individual behaviour, there is still a lack of understanding regarding the complex cumulative ways in which social practices might be shifted, which would involve tackling the different dimensions of a practice – from images and meanings, to people and their competencies, and finally the material dimension of consumption. In relation to food consumption, these might involve meanings around festive or holiday meals, the competencies to prepare such a meal, and the space and access to products that make preparing and sharing a meal possible. Sahakian and Wilhite (2014) found that at least two of these elements needed to change in order to engender a rupture in routines or habits related to preparing and eating a meal; in the example above, opting for a vegetarian meal for Thanksgiving (instead of turkey) could indicate that traditional meanings around meat as 'social food' is shifting to meanings regarding moral questions of animal welfare.

In another study (Plessz et al., 2016), food consumption is seen as being subject to socially constructed guidelines about what ought or should be consumed, what the authors term 'prescriptions'; life-course events such as moving in with a partner were found to have an important role to play in how people take up prescriptions. The dynamic relation between life-course and everyday consumption practices is seen as a key area for encouraging more sustainable consumption (Schäfer and Jaeger-Erben, 2012; Rau, Davies, and Fahy, 2014; Greene and Rau, 2018).

In the course of this project, we found that time is an essential resource for enacting specific prescriptions, and that a lack of temporal resources might lead to tradeoffs – more than a lack of financial resources, for example. Given time constraints, mobility and transit have a structuring effect on what food can be easily accessed, where and when. The availability of food retailers and products on the way between work and home, among others, seem to influence food purchases. In addition to time, social dynamics inside and outside the home are a defining feature for taking up prescriptions and changing practices. The composition of the household, most importantly the presence of children, is a central element for the adoption food prescriptions. Outside the home, people carry and disseminate prescriptions, and one person or one household diet is linked to its relationships and group of peers (Godin and Sahakian, 2018).

Overcoming the challenges: an interdisciplinary approach

In the following section, we describe how our research design and methodological choices helped us overcome the challenge of integrating social practice theory approaches from the sociology of consumption, with the LCA framework from environmental studies. Some of our choices relate to the interdisciplinary aspect of our methodology, while others are prompted by the possibilities and constraints brought about by our object of study – healthy and sustainable food consumption.

Interdisciplinarity as an answer to a complex object and question

Interdisciplinarity is gaining more importance in academic life, quickly becoming a central feature of institutional organisation and funding schemes at the national and international level. Despite this, there is no consensual definition of interdisciplinarity, and its meaning changes along with funding bodies and modes of academic governance (Cooper, 2013). In this project, interdisciplinarity involves a shared view of the problem among team members from different backgrounds, a common language, and consensus building at the stage of research design and implementation. Stakeholders and practitioners are involved at each step of the project, which brings us close to trans-disciplinarity as defined by Lang et al. (2012).

Interdisciplinarity is ingrained in sustainability science: given the multiple dimensions of sustainability as an object of study, some argue that progress in the development of knowledge 'will require fostering problem-driven, interdisciplinary research' (Kates et al., 2001, p. 641). In this spirit, climate change is often presented as a domain of scientific enquiry that requires interdisciplinary approaches (Cooper, 2013) and used as an example to study their implementation (e.g. Castán Broto et al., 2009). The alliance between disciplines pervasive to the study of sustainability means that novel methodological, conceptual and epistemological challenges have to be tackled in order to shed light on complex, new problems (Kates et al., 2001). For example, interdisciplinary approaches integrating social practice theory and material flow analysis tools have been successful in understanding domestic food waste by looking simultaneously at the household metabolism, including the quantity of waste, and at consumption practices revealing why and in what way food ends up being thrown away (Leray et al., 2016).

To begin the interdisciplinary work in this project, we developed research questions that would lead to synergies between our respective disciplinary understandings of food and diets: how do prescriptions and practices evolve around what are considered as 'healthy and sustainable' diets? How can the health *and* environmental impacts of dietary scenarios be assessed? How can research support transitions towards healthier and more sustainable diets in Switzerland? Answering those questions meant bringing together food prescriptions and practices, from sociology, with a novel LCA framework that combines environmental and health benefits and impacts of various dietary scenarios, anchored in ecological economics, in a first stage. The second stage, integrating new knowledge into a transition management perspective, has yet to begin at the time of writing.

The proposed questions aimed to combine methods towards streamlined goals, which are a defining feature of our project. The goals to be fulfilled through qualitative research and social practice approaches are controlled by the limitations in life cycle assessment and data availability (e.g. there is not always 'local' data available on the environmental impacts of food production and the system boundaries of what is 'local' are difficult to define), while the environmental and health impact analysis is also controlled by the outcome of the sociological research with respect to which dietary prescriptions to assess. In addition to the collaboration of the research team, a variety of non-academic stakeholders were involved in an advisory group providing guidance for selecting focus areas, in an attempt to ensure the relevance of our project for social and political actors, and for broader society. In this spirit, the LCA results will also be fed into workshops towards designing transitions in dietary changes at a later stage, thus informing qualitative and participatory methods with quantitative research results.

A social practice theory approach to food and diets

The first stage of our collaboration involved the identification of the most significant prescribed so-called healthy and/or sustainable diets in Switzerland. Based on the work of Plessz et al. (2016), we understand prescriptions as discourses stating what and how it is best to eat, designed to influence practices and providing a framework for conduct. Prescriptions can, for example, take the form of official nutrition guides, such as the Swiss food pyramid; they can be general principles carried by various stakeholders, such as the prescription for organic food consumption; and they are often the object of heated debates, as is the case for gluten-free diets without a diagnosed medical condition. We consider prescriptions as a resource in the establishment of practices, along with material resources, and competencies and skills, among others (Halkier, 2009; Plessz et al., 2016; Warde, 2013). To identify dominant food prescriptions in Switzerland, we conducted interviews with practitioners working for organisations interested in food consumption (five in-depth interviews), completed an institutional mapping of the actors involved in the promotion of healthy and/or sustainable food consumption (90 institutional actors), studied media discourses on food and eating (188 Swiss newspaper articles and issues of 8 magazines, in French and in German) and engaged in participant observation in events (five events), such as policy consultations, relevant to our research area.

We uncovered several emerging and established prescriptions for dietary practices formulated and carried out by public authorities, health agencies, non-governmental organisations, schools and workplaces, economic actors such as participants in community-supported agriculture or retailers, health professionals, traditional media and the network of peers, among others. Five prescriptions we identified have an important impact on how the LCA and health impact assessment are conducted. First, the ideal of a balanced diet is shared by most people, and is closely linked to national dietary guidelines which, in the case of Switzerland, take the form of the Swiss food pyramid (Société Suisse de Nutrition [SSN], 2018). Next there are 'local and seasonal' diets, often associated with the consumption of 'natural and organic' food as another dietary prescription, but nonetheless distinct. Meat consumption proves to be quite divisive, with the supporters of vegetarian and vegan diets on one side, and on the other side people who advocate 'less, but better' meat consumption, although there are distinct subcategories to each prescription. Our fieldwork showed that prescriptions on meat consumption are not on a continuum but are two opposing views: on one side are the people who consider killing animals as immoral, on the other side, those who do not have an issue with the killing of animals, but insist on the importance of the well-being of the cattle, in life and death. Other prescriptions include slimming diets, proscriptive diets excluding of one or many kinds of foods, and body-oriented diets, such as 'detox' diets or clean eating, designed to 'clean' or avoid polluting the body (Godin and Sahakian, 2018).

It is worth noting that, while all prescriptions can be seen as relating to health, only a subset relates to environmental sustainability. The qualitative research showed that in consumers' representations, both categories seem to be conflated: sustainable food consumption is seen as healthier, while a healthy diet is seen as more sustainable. Moreover, health seems to have much more traction when it comes to influencing practices, and is often presented as the main reason for engaging in 'sustainable' practices such as local food consumption. Assumptions regarding the 'healthiness' of local and seasonal, but also natural and organic, food products are for the most part a matter of trust towards the producers as well as the retailer. Trust grows stronger when there is a direct contact with one or the other.

With a better grasp of food prescriptions around health and sustainability in Switzerland, we looked to understand their expression in everyday consumption practices. We conducted participant observation and short, open-ended interviews with employees and consumers in a supermarket in Lausanne (2 days, 11 interviews), which set the stage for a series of semi-directed interviews with consumers (9 interviews with 10 participants), as well as focus group discussions (5 focus groups with a total of 29 participants). We built the sampling for consumer interviews in order to access different types of household (one-person household, nuclear families, couple without children, etc.) with members at various life stages (students, young parents, retirees, etc.). Based on a social practice approach, our goal was to understand which social institutions, dimensions of everyday life and constraints proper to the different life stages contribute the most to organising food consumption practices and the adoption of dietary prescriptions. For focus groups, we gathered participants with strong and possibly controversial opinions on food and eating in order to see how different prescriptions interact at the discursive level, what kinds of tensions exist between them and how these tensions are resolved or not in practices. During interviews and focus groups, we used visual scenarios and pictures to stimulate discussions and bring the different representations of food consumption to the surface, in a method known as photo-elicitation (Harper, 2002; Lachal et al., 2012; Meyer, 2017).

The qualitative research allowed us to have a closer look not only at consumers' representations of the different prescriptions but also at the elements at play when prescriptions are put into practice. They can be of individual nature (e.g. life stage, social network, competencies and skills), they can relate to the socio-cultural dimension (e.g. social norms, traditions, collective identity), or be linked to material limitations (e.g. accessibility of products, mobility, available tools and appliances). Different combinations of these elements can either facilitate or impede the adoption, voluntary or not, of specific prescriptions, or lead to their distortion.

Prescriptions, consumers' representations and practices and other consulted stakeholders (e.g. policy-makers) served to identify a number of variables that people associated with the healthiness and sustainability of Swiss diets. Among these variables, the most important for consumers were local food production, seasonal food consumption, organic food production, meat consumption and the level of food processing – all elements that are tangible for consumers (e.g. related to product labels). However, the variables identified through qualitative research do not always align with the quantitative findings from LCA research. For example, previous LCAs of food systems generally demonstrate that agricultural production (and, for example, fertilisers) is more influential on environmental impact of the food item than the transport of the food across far distances.

Furthermore, consumers' idea of what is 'local' or 'regional' may not include foods coming from regions in neighbouring countries (such as northern Italy) where transport distance may actually be similar to Swiss foods. The idea of 'local' or 'regional' does not include the type of transport used (e.g. train, lorry) which also has a large influence on the impact. Additionally, although consumers may associate 'organic' with more 'sustainable', quantitative research demonstrates trade-offs such that organic production has less pesticide input but can require more land (Roy et al., 2009). Finally, qualitative research may indicate 'less meat' as an important association with health and sustainability, but quantitative research shows large differences among the impacts (both on health and the environment) across different meat types, as well as tradeoffs (e.g. processed meat has a higher health impact but can have lower environmental impacts than non-processed meat).

Articulating environmental LCA and health impact assessment

In the second stage of this project, we aimed at drawing a set of diets that depicts accurately what people eat, in order to estimate their health and environmental impact. Our approach takes a dietary perspective, where one single product does not contribute to 'health', but the overall dietary consumption protects health or increases disease risk. Using the Swiss Dietary Survey 'MenuCH' (Bochud et al., 2017), the first national data on food consumption based on memory recall surveys and the visual identification of portion size through photographs, we matched the hundreds of food items available in MenuCH to existing environmental data. We also considered seasonal production, processing, region of production and packaging as additional variables meaningful to consumers, as emerged from the qualitative work.

An environmental life cycle assessment (LCA) framework was used to assess the environmental impacts of the Swiss diet. LCA is a framework that involves defining a system of processes and quantifying the material and energy flows from these processes by collecting life cycle inventory (LCI) – for example, from established databases. The defined system considers relevant 'life cycle' stages from extraction and production, to consumption and final waste. LCI was then translated into impacts (e.g. climate change) through an environmental life cycle impact assessment (LCIA) method. LCIA methods do not typically include the human health impacts related to food consumption or product use, however, the Global Burden of Disease study series (GBD, 2015; Risk Factors Collaborators, 2016) shows that when combined, dietary risk factors are a leading cause of death globally. The Global Burden of Disease synthesises the published literature on health outcomes to date to better understand and compare what causes death and disease in different populations. It also provides the data needed to consider risk factors for a population (e.g. above or below which level of consumption leads to an increased risk of certain diseases). Ultimately our aim is to apply a novel

LCA framework that includes Global Burden of Disease data on dietary health impacts such as that done previously in a proof-of-concept study by Stylianou et al. (2016). Using this combined LCA and health impact assessment framework, we hope to assess what people eat on average in Switzerland and relate to different prescriptions. Through this work we can identify tradeoffs or misconceptions between various so-called healthy and sustainable practices – for example, where the perception of local food being more sustainable is not always reflected by the life cycle-based impact assessment approach.

Challenges of interdisciplinarity

For this project, interdisciplinarity involves a shared view of 'healthy and sustainable' food consumption among team members from different backgrounds and a strong coordination between all involved, to ensure our respective approaches and methodologies work well together. Figure 3.1 illustrates the integration of LCA and social practice theory approaches. When assessing dietary scenarios, LCA practitioners typically start from hypothetical diets that are constructed from available knowledge (e.g. from food pyramids, from a modelled diet not containing meat and sometimes from actual dietary survey data). This approach is disjointed from consumer representations and not always useful to relate back to most people's beliefs and practices. In this research project, consumers' practices and representations studied through the lens of social practice theory inform different dietary scenarios for LCA, to better understand what healthy and sustainable diets are in Switzerland and, as a next step, to support transition towards healthier and more sustainable diets in terms of actual consumption.

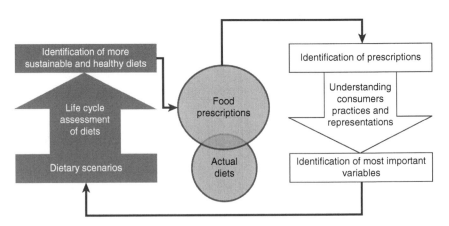

Figure 3.1 Combining LCA and social practice approaches to contribute new knowledge on 'sustainable and healthy' diets.

The first challenge we met was to build a common vocabulary that would work for all of us. Second came the problem of ambiguous categories that emerged of the qualitative fieldwork and posed difficulties for LCA and health impact assessment. A third difficulty lies in the necessity of putting aside disciplinary baggage. Moving from practices to volumes and from meanings to nutrients means negotiating, compromising and dealing with tradeoffs so that we can take full advantage of the articulation of our respective views of healthy and sustainable food consumption.

Engaging in a pedagogical effort

One of the first steps we took at the beginning of the project was to develop a common vocabulary by creating a glossary to define notions that were meaningful to the different disciplinary fields. This simple exercise shed light on the need for clear definitions and the way they contribute to framing the common understanding of the project and its object, in this case healthy and sustainable Swiss diets. For example, the sociologists on the team pushed to formulate the problem in terms of practices and representations that account for both the individual and social dimensions of a phenomenon, as opposed to individual behaviour and perceptions, a formulation at risk of leading us into a definition of the problem that would obscure its social dimension, falling into the trap of the individualisation mentioned earlier.

Through the development of a common vocabulary, each team member had to engage in a pedagogical effort. Indeed, this exercise sent all of us back to the fundamentals of our disciplines, having to navigate between different views of the same issue. For the sociologists, it meant learning about the basics of LCA and the way the calculation of the environmental impact of a given product is made, as well as the criterion, variables and measurements guiding the health impact assessment at the nutritional level. It also meant defining concepts that are otherwise part of the basic toolkit and usually taken for granted, such as discourse or representation. For example, much time was spent describing the social embeddedness of everyday life, which guided the qualitative research. In this project, we aim at identifying strategies to induce a transformation of practices without necessarily appealing to individual rationality or motivation, nor veering towards structuralism. To this end, we worked to explain how everyday practices are dependent on the food system as a whole, as well as social norms and expectations around food in given contexts, and how such insights can serve to better understand what transformations might be possible towards 'sustainable and healthy' food consumption pathways. This meant going back to the core project of sociology and inviting all partners to develop their sociological imagination, to see how the actions of one person are tied to larger social dynamics, thus taking a problem usually framed as an individual one and looking at it from a *systemic and interrelated* standpoint – to use the language more common in environmental studies.

Working with ambiguous categories

Given the sequential structure of our methodology, with sociological research done first, a crucial issue was to make the categories that emerged from the qualitative fieldwork work for everyone. At this point, the discrepancies between consumers' and stakeholders' representations, the sociological definitions, and the needs of LCA and health impact assessment became highly visible. Constrained by the fundamentally different nature of our respective disciplines, we had to engage in a negotiation process that meant for each party to accept compromises in order to build categories meaningful for all of us and the project goals.

The choice of words and the definition of the notions they denominate carried minimal impact when mobilised by one discipline much more than by the other. This was the case for the concepts of practices and representations, as opposed to behaviour and perception. The use of these terms created the need for some explanation from the sociologists' part, but was not disputed as they had strong implications for only one discipline. For other, more fundamental categories, such as the notion of health, we worked in parallel. The sociologists mostly used the lay understanding of the notion, in which health can mean maintaining a body free of pollutants, or accessing all the vitamins and nutrients necessary to maintain general well-being, for example. Health impact assessments, on the other hand, generally leave aside any subjective understanding of health in favour of scientific evidence on health outcomes and disease burden measured in disability adjusted life years (DALYs), which represent the amount of healthy life lost in a population due to death and disease (GBD, 2015; Risk Factors Collaborators, 2016). That being said, the environmental scientists on the team recognised that health could not be limited to products or items within a dietary basket, but rather would reflect overall food consumption and related habits – such as sport activity, for example. In this respect, our understanding of a healthy and sustainable *diet* was very much aligned.

The discussions about the examples mentioned happened without much friction. Some other categories, however, proved more complicated, especially when the lay perspective, sociological understanding and both LCA and health impact assessment needed to find some common ground. 'Local' or 'regional' food consumption is the best example for such categories, as the impossibility to achieve a definition that could mirror consumers' representations while being precise enough to conduct environmental assessment threatened the production of knowledge that should be part of the outcome of our project. Eriksen (2013, p. 47) rightly notes,

> Perceptions of local food vary, for example, with the location of the consumer. To some it refers to food that has been produced in the locality close to where 'I' live. To others food is considered local if it is produced in the same country in which it is consumed. There is also great variability in what constitutes local food for producers and for consumers.

In Switzerland, local or regional food is often defined by consumers through cantonal borders, although some institutions, retailers and labels might rely on distance in terms of kilometres to identify such products. At the same time, our research shows that local food consumption is seen as being healthier based on the consumer's perception that the producers they can talk with use fewer pesticides and antibiotics, for example. Local consumption is also seen as more socially responsible and environment friendly. However, this lack of a consistent definition of what would be a 'local' or 'regional' product is problematic when attempting to assess the environmental impacts of consuming such foods, as consumers' representations and their translation in objective measurements (e.g. food miles) differ depending on their geographic zone. In other words, LCA can be used to assess the impact of food distribution, but often uses assumptions regarding transport distances which may not be aligned with a consumer's perception of 'local' or regional, and these assumptions are disjointed from what individual consumers perceive as local.

Putting aside disciplinary baggage

Social practice theory approaches led to the identification and description of prescriptions around healthy and sustainable diets in Switzerland, along with a deeper understanding of key elements that allow or hinder the adoption of related practices at the individual and household levels. LCA results should be viewed as a screening and prioritisation information that can help indicate environmental impacts across life cycles of food items in a consistent and quantitative way. Bringing our results together provides some elements for a common view of how to support transitions towards healthier and more sustainable Swiss diets, but the problem of achieving a tight integration in designing ways of achieving our common, normative goal, which is essential if we are to truly make the best of our collaboration, is still to be tackled.

At the time of writing, we have developed a methodology for assessing food consumption from a social practice perspective, while assessing healthy and sustainable diets with LCA. What remains is to bring our findings together, and better understand the implications for transitions towards 'healthy and sustainable' diets in Switzerland, the last phase of our project. From a social practice theory perspective, the most important issue for LCA and health impact analysis is to move away from unitary analysis back to practices and everyday life. While new knowledge on impacts is important, we shy away from being prescriptive and look to better understand tradeoffs from the different diets represented as 'healthy and sustainable'. Social practice theory approaches point out the most important elements for consumers and policy-makers and the dynamics behind them, which might not relate to the most relevant findings and categories for environmental and health impact assessment.

Addressing this issue is critical to the success of this project, but it takes tremendous work and commitment to put aside disciplinary baggage, as well as people who are willing to take on different perspectives. In this context, the reality of precarious academic work can be disruptive, as people tend to change over time – even on projects conducted on a one- or two-year time span. Given institutional, interpersonal and disciplinary dynamics, to make the best of inter-disciplinary collaborations, coordination is key: to defuse conflicts or avoid them altogether, but also to define common goals that fulfil the academic and institutional requirements of all partners.

Concluding remarks: managing tradeoffs in interdisciplinary approaches

The research project is ongoing, and final results as to environmental and health impacts as well as on possible transitions are still pending. That being said, this chapter aims towards uncovering how challenges in the study of healthy and sustainable diets might be overcome through interdisciplinary approaches, and what approaches might be combined towards the normative goal of achieving 'healthier and more sustainable diets' among the Swiss population.

In our explicit focus on prescriptions, we aimed to uncover what everyday people think they ought to eat when it comes to healthy and sustainable diets. There is an implicit understanding in this research project that the environmental and health impacts we will assess in relation to different diets will give us better knowledge of the most important prescriptions, or at least scientific knowledge in relation to a variety of environmental indicators along with the Global Burden of Disease studies. However, the combination of our two disciplines can point to a general direction, but does not provide solutions for a transition towards health-ier and more sustainable diets. It is more likely that the LCA results will not provide one answer for how to improve diets, but rather an assessment of trade-offs in recognising the health and environmental value of certain diets, across provisioning systems and in relation to how food is produced, distributed, pack-aged and consumed. For example, we might be able to quantify the health and environmental tradeoffs associated with eating a diet that includes imported so-called superfoods, such as avocado. Perhaps the health benefits will be revealed, in relation to consuming more fruits, but there may be an environmental impact tradeoff – for example, if the avocado is flown from overseas. This leaves many questions open as to which prescriptions should be put forward that can be shared by prescribers as diverse as national health agencies, schools, community organi-sations, retailers, friends, families or doctors.

Our interdisciplinary approach to food consumption through the integration of social practice theory and LCA allowed us to give a sound scientific basis to the qualitative findings, which for a big part rely on a comprehensive approach and subjective construction of the relevant categories. Such a strategy can serve to address the concerns of stakeholders and, to some extent, consumers, which is part of the added value of our project. At the same time, the LCA work can be improved by being informed by what is meaningful to individuals, communities and institutional stakeholders, in relation to prescriptions and practices. Our collaboration, however, did not aim to produce larger theoretical or conceptual transformations. Rather, it is a first step to open the conversation between LCA and social constructs – for example, to inform which variables are used for the calculation – and are also relevant to the average person and their decision-making process (Goldstein et al., 2016).

Ultimately, the study of healthy and sustainable food is all about tradeoffs: not only in relation to the types of solutions that might be proposed towards dietary transitions but also between disciplines and approaches. This research project is an attempt to grapple with complexity in earnest, where compromises are better than solely a disciplinary approach to food consumption.

Acknowledgements

This chapter is based on a research project funded by the Swiss National Science Foundation (SNSF) under grant number 406940_166763 in the frame of the national research programme 'Healthy nutrition and sustainable food production' (NRP 69), coordinated by Suren Erkman (University of Lausanne) and co-coordinated by Marlyne Sahakian (University of Geneva) and Claudia Binder (EPFL), with the collaboration of Olivier Jolliet (University of Michigan). We gratefully acknowledge the members of our advisory committee for their input and all of the people who agreed to participate in our study.

Note

1 Project 'Tipping points toward healthy and sustainable Swiss diets: Assessing prescriptions, practices, and impacts', PNR69, Swiss Nation Science Foundation.

References

Antonetti, P. and Maklan, S. (2014). Feelings that make a difference: How guilt and pride convince consumers of the effectiveness of sustainable consumption choices. *Journal of Business Ethics*, 124(1), pp. 117–134. https://doi.org/10.1007/s10551-013-1841-9

Blake, J. (1999). Overcoming the value-action gap in environmental policy: Tensions between national policy and local experience. *Local Environment: The International Journal of Justice and Sustainability*, 4(3), pp. 257–278. https://doi.org/10.1080/1354983 9908725599

Bochud, M., Chatelan, A., Blanco, J.-M., et al. (2017). *Anthropometric characteristics and indicators of eating and physical activity behaviors in the Swiss adult population. Results from menuCH 2014–2015*. Federal Food Safety and Veterinary Office and Federal Office of Public Health FOPH, Bern, Swiss Confederation.

Born, B. and Purcell, M. (2006). Avoiding the local trap: Scale and food systems in planning research. *Journal of Planning Education and Research*, 26(2), pp. 195–207. https://doi.org/10.1177/0739456X06291389

Bourdieu, P. (1979). *La distinction: Critique sociale du jugement*. Paris: Minuit.

Campbell, B., Beare, D., Bennett, E., et al. (2017). Agriculture production as a major driver of the Earth system exceeding planetary boundaries. *Ecology & Society*, 22. https://doi.org/10.5751/ES-09595-220408

Castán Broto, V., Gislason, M. and Ehlers, M.-H. (2009). Practising interdisciplinarity in the interplay between disciplines: Experiences of established researchers. *Environmental Science & Policy*, 12(7), pp. 922–933. https://doi.org/10.1016/j.envsci.2009.04.005

Chastonay, P., Simos, J., Cantoreggi, N., et al. (2017). Health priorities in French-speaking Swiss cantons. *International Journal of Health Policy and Management*, 7(1), pp. 10–14. https://doi.org/10.15171/ijhpm.2017.91

Cohen, M.J. and Murphy, J. (eds.) (2001). *Exploring sustainable consumption: Environmental policy and the social sciences*. Oxford: Elsevier.

Cooper, G. (2013). A disciplinary matter: Critical sociology, academic governance and interdisciplinarity. *Sociology*, 47(1), pp. 74-89. https://doi.org/10.1177/0038038512444812

Dubuisson-Quellier, S. and Plessz, M. (2013). La théorie des pratiques. Quels apports pour l'étude sociologique de la consommation? *Sociologie*, 4(4). Available at: http://sociologie.revues.org/2030

Eriksen, S.N. (2013). Defining local food: Constructing a new taxonomy – three domains of proximity. *Acta Agriculturae Scandinavica, Section B – Soil & Plant Science*, 63(Suppl. 1), pp. 47-55. https://doi.org/10.1080/09064710.2013.789123

Ernstoff, A., Stylianou, K.S. and Goldstein, B. (2017). Response to: Dietary strategies to reduce environmental impact must be nutritionally complete. *Journal of Cleaner Production*, 162, pp. 568–570. https://doi.org/10.1016/j.jclepro.2017.05.205

Eshel, G., Shepon, A., Noor, E., et al. (2016). Environmentally optimal, nutritionally aware beef replacement plant-based diets. *Environmental Science & Technology*, 50, pp. 8164–8168. https://doi.org/10.1021/acs.est.6b01006

Fahy, F. and Rau, H. (eds.) (2013). *Methods of sustainability research in the social sciences*. London: Sage.

FAO (2012). *Guidance note: Integrating the right to adequate food into food and nutrition security programmes*. Available at: www.fao.org/docrep/017/i3154e/i3154e.pdf

FCRN (Mar. 26, 2018). *Home*. Available at: www.fcrn.org.uk

Fern, E.B., Watzke, H., Barclay, D.V., et al. (2015). The nutrient balance concept: A new quality metric for composite meals and diets. *PLoS ONE*, 10, p. e0130491. https://doi.org/10.1371/journal.pone.0130491

FOPH (2012). *Swiss nutrition policy 2013–2016*. Available at: https://extranet.who.int/nutrition/gina/sites/default/files/CHE%202013-2016%20Swiss%20Nutrition%20Policy%20EN.pdf [Accessed 14 Apr. 2018].

FSO (2015). *Swiss agriculture – pocket statistics 2015*. Available at: www.bfs.admin.ch/bfsstatic/dam/assets/349914/master [Accessed 31 Aug. 2018].

Galani, C., Schneider, H. and Ruten, F.F.H. (2007). Modelling the lifetime costs and health effects of lifestyle intervention in the prevention and treatment of obesity in Switzerland. *International Journal of Public Health*, 52(6), pp. 372–382. https://doi.org/10.1007/s00038-007-7014-9

GBD 2015 Risk Factors Collaborators (2016). Global, regional, and national comparative risk assessment of 79 behavioural, environmental and occupational, and metabolic risks or clusters of risks, 1990-2015: A systematic analysis for the Global Burden of Disease Study 2015. *The Lancet*, 388(10053), pp. 1659–1724.

Giddens, A. (1984). *The constitution of society: Outline of the theory of structuration*. Berkeley: University of California Press.

Godin, L. and Sahakian, M. (2018). Cutting through conflicting prescriptions: How guidelines inform healthy and sustainable diets in Switzerland. *Appetite*, In Press. https://doi.org/10.1016/j.appet.2018.08.004

Goldstein, B., Hansen, S.F., Gjerris, M., et al. (2016). Ethical aspects of life cycle assessments of diets. *Food Policy*, 59, pp. 139–151. https://doi.org/10.1016/j.foodpol.2016.01.006

Greene, M. and Rau, H. (2018). Moving across the life course: A biographic approach to researching dynamics of everyday mobility practices. *Journal of Consumer Culture*, 18(1), pp. 60–82. https://doi.org/10.1177/1469540516634417

Halkier, B. (2009). Suitable cooking? Performances and positioning in cooking practices among Danish women. *Food, Culture & Society*, 12(3), pp. 357–377. https://doi.org/10.2752/175174409X432030

Halkier, B. and Jensen, I. (2011). Doing "healthier" food in everyday life? A qualitative study of how Pakistani Danes handle nutritional communication. *Critical Public Health*, 21(4), pp. 471–483. https://doi.org/10.1080/09581596.2011.594873

Hallström, E., Carlsson-Kanyama, A. and Börjesson, P. (2015). Environmental impact of dietary change: A systematic review. *Journal of Cleaner Production*, 91, pp. 1–11. https://doi.org/10.1016/j.jclepro.2014.12.008

Harper, D. (2002). Talking about pictures: A case for photo elicitation. *Visual Studies*, 17(1), pp. 13-26. https://doi.org/10.1080/14725860220137345

Heller, M.C. and Keoleian, G.A. (2015). Greenhouse gas emission estimates of U.S. dietary choices and food loss. *Journal of Industrial Ecology*, 19(3), pp. 391–401. https://doi.org/10.1111/jiec.12174

Heller, M.C., Keoleian, G.A. and Willett, W.C. (2013). Towards a life cycle-based, diet-level framework for food environmental impact and nutritional quality assessment: A critical review. *Environmental Science & Technology*, 47, pp. 12632–12647. https://doi.org/10.1021/es4025113

Jaeger-Erben, M. and Offenberger, U. (2014). A practice theory approach to sustainable consumption. *GAIA – Ecological Perspectives for Science and Society*, 23(1), pp. 166–174. https://doi.org/10.14512/gaia.23.S1.4

Jolliet, O., Saadé-Sbeih, M., Shaked, S., et al. (2015). *Environmental life cycle assessment*. Boca Raton: CRC Press.

Kamm, A., Hildesheimer, G., Bernold, E., et al. (2015). *Ernährung und Nachhaltigkeit in der Schweiz: Eine verhaltensökonomische Studie*. BAFU: FehrAdvice & Partners AG.

Kates, R.W., Clark, W.C., Corell, R., et al. (2001). Sustainability science. *Science*, 292(5517), pp. 641-642. www.jstor.org/stable/3083523

Kollmuss, A. and Agyeman, J. (2002). Mind the gap: Why do people act environmentally and what are the barriers to proenvironmental behavior? *Environmental Education Research*, 8(3), pp. 239–260. https://doi.org/10.1080/13504620220145401

Lachal, J., Speranza, J., Taïeb, O., et al. (2012). Qualitative research using photo-elicitation to explore the role of food in family relationships among obese adolescents. *Appetite*, 58(3), pp. 1099–1105. https://doi.org/10.1016/j.appet.2012.02.045

Lang, D.J., Wiek, A., Bergmann, M., et al. (2012). Transdisciplinary research in sustainability science: Practice, principles, and challenges. *Sustainability Science*, 7(1), pp. 25–43. https://doi.org/10.1007/s11625-011-0149-x.

Leray, L., Sahakian, M. and Erkman, S. (2016). Understanding household food metabolism: Relating micro-level material flow analysis to consumption practices. *Journal of Cleaner Production*, 125, pp. 44–55. https://doi.org/10.1016/j.jclepro.2016.03.055.

Masset, G., Soler, L.-G., Vieux, F., et al. (2014). Identifying sustainable foods: The relationship between environmental impact, nutritional quality, and prices of foods representative of the French diet. *Journal of the Academy of Nutrition and Dietetics*, 114(6), pp. 862–869. https://doi.org/10.1016/j.jand.2014.02.002

Meyer, M. (2017). La force (é)vocative des archives visuelles dans la situation d'enquête par entretiens: Une étude par photo-élicitation dans le monde ambulancier. *Revue Française des Méthodes Visuelles*, 1. Available at: https://rfmv.fr

Nemecek, T., Jungbluth, N., Canals, L.M.I., et al. (2016). Environmental impacts of food consumption and nutrition: Where are we and what is next? *International Journal of Life Cycle Assessment*, 21, pp. 607–620. https://doi.org/10.1007/s11367-016-1071-3

Notarnicola, B., Tassielli, G., Renzulli, P.A., et al. (2017). Environmental impacts of food consumption in Europe. *Journal of Cleaner Production*, Towards eco-efficient agriculture and food systems: Selected papers addressing the global challenges for food systems, including those presented at the Conference "LCA for feeding the planet and energy for life" (6–8 Oct. 2015, Stresa & Milan Expo, Italy), 140, pp. 753–765. https://doi.org/10.1016/j.jclepro.2016.06.080

OECD (2011). *Switzerland – agricultural policy monitoring and evaluation 2011*. Available at: www.oecd.org/switzerland/switzerland-agriculturalpolicymonitoringandevaluation2011.htm [Accessed 31 Aug. 2018].

Plessz, M., Dubuisson-Quellier, S., Gojard, S., et al. (2016). How consumption prescriptions affect food practices: Assessing the roles of household resources and life-course events. *Journal of Consumer Culture*, 16(1), pp. 101–123. https://doi.org/10.1177/1469540514521077

Rau, H., Davies, A. and Fahy, F. (2014). Conclusion: Moving on. Promising pathways to a more sustainable future. In: A. Davies, F. Fahy and H. Rau, eds., *Challenging consumption: Pathways to a more sustainable future*. London: Routledge.

Rockström, J., Stordalen, G.A. and Horton, R. (2016). Acting in the anthropocene: The EAT-Lancet commission. *The Lancet*, 387(10036), pp. 2364–2365. https://doi.org/10.1016/S0140-6736(16)30681-X

Roy, P., Nei, D., Orikasa, T., et al. (2009). A review of life cycle assessment (LCA) on some food products. *Journal of Food Engineering*, 90, pp. 1–10. https://doi.org/10.1016/j.jfoodeng.2008.06.016

Sahakian, M. and Wilhite, H. (2014). Making practice theory practicable: Towards more sustainable forms of consumption. *Journal of Consumer Culture*, 14(1), pp. 25–44. https://doi.org/10.1177/1469540513505607

Schäfer, M. and Jaeger-Erben, M. (2012). Life events as windows of opportunity for changing towards sustainable consumption pattern? The change in everyday routines in life course transitions. In: R. Defila, A.D. Giulio and R. Kaufmann-Hayoz, eds., *The nature of sustainable consumption and how to achieve it: Results from the focal topic "From knowledge to action – new paths towards sustainable consumption"*. Munich: oekom.

Schatzki, T.R. (1996). *Social practices: A Wittgensteinian approach to human activity and the social*. Cambridge: Cambridge University Press.

Shove, E. (2010). Beyond the ABC: Climate change policy and theories of social change. *Environment and Planning A*, 42, pp. 1273–1285. https://doi.org/10.1068/a42282

Shove, E. and Pantzar, M. (2005). Consumers, producers and practices: Understanding the invention and reinvention of Nordic walking. *Journal of Consumer Culture*, 5(1), pp. 43–64. https://doi.org/10.1177/1469540505049846

Shove, E., Pantzar, M. and Watson, M. (2012). *The dynamics of social practice: Everyday life and how it changes.* London: Sage.

SNSF (Mar. 26, 2018). *NRP 69 healthy nutrition and sustainable food production.* Available at: www.snf.ch/en/researchinfocus/nrp/nrp69-healthy-nutrition-and-sustainable-food-production

SSN (Apr. 12, 2018). *Swiss food pyramid.* Available at: www.sge-ssn.ch/media/sge_pyramid_E_basic_20161.pdf

Stylianou, K.S., Heller, M.C., Fulgoni, V.L. III, et al. (2016). A life cycle assessment framework combining nutritional and environmental health impacts of diet: A case study on milk. *International Journal of Life Cycle Assessment*, 21(5), pp. 734–746. https://doi.org/10.1007/s11367-015-0961-0

Tilman, D. and Clark, M. (2014). Global diets link environmental sustainability and human health. *Nature*, 515, pp. 518–522. https://doi.org/10.1038/nature13959

Tobler, C., Visschers, V.H.M. and Siegrist, M. (2011). Eating green. Consumers' willingness to adopt ecological food consumption behaviors. *Appetite*, 57(3), pp. 674-682. https://doi.org/10.1016/j.appet.2011.08.010

Tukker, A., Huppes, G., Suh, S., Heijungs, R., Guinée, J., Koning, A.D., . . . Nielsen, P. (2006). *Environmental impacts of products.* Seville, Spain: ESTO/IPTS.

Walker, C., Gibney, E.R. and Hellweg, S. (2018). Comparison of environmental impact and nutritional quality among a European sample population – findings from the Food4Me study. *Scientific Reports*, 8, p. 2330. https://doi.org/10.1038/s41598-018-20391-4

Warde, A. (2013). What sort of practice is eating? In: E. Shove and N. Spurling, eds., *Sustainable practices: Social theory and climate change.* London: Routledge.

WHO (June, 2017). *Noncommunicable diseases.* Available at: www.who.int/mediacentre/factsheets/fs355/en/

Willett, W., Rockström, J., Loken, B., Springmann, M., Lang, T., Vermeulen, S., . . . Murray, C. J. L. (2019). Food in the Anthropocene : The EAT–Lancet Commission on healthy diets from sustainable food systems. *The Lancet*, 393(10170), pp. 447–492. https://doi.org/10.1016/S0140-6736(18)31788-4

Chapter 4

Traditional or cultural relativist school meals?

The construction of religiously sanctioned school meals on social media

Christine Persson Osowski
and Ylva Mattsson Sydner

Introduction

As a result of migration, Sweden can today be labelled a culturally and religiously diverse society, as many other western countries. However, Sweden is different from other multicultural societies in that it found itself at the very extreme position in the World Values Survey for secular-rational values among the population, with a low focus on religion and self-expression values, including tolerance of foreigners and equality (Institute for Comparative Survey Research, 2017). At the same time, Sweden as a society is characterised as being a homogenous and Protestant Scandinavian welfare state. The process of migration has presented the Swedish welfare state with new challenges due to the need of welfare services for a diverse heterogenic population. This diversity encompasses different religious aspects, and thereby religion in the public sphere is visible in a new way. One example of these challenges and how religion has been visible is school meals, which are served to all pupils in primary school every day. As migration has increased in the past few years, the demand for religiously sanctioned school meals has become more pronounced. This has produced a discussion, and there are many different opinions among laypeople with regard to religion in the public sphere. One issue that is currently being discussed is whether religiously sanctioned school meals should be served or not. This chapter focuses on this ongoing debate on social media about religiously sanctioned school meals.

Traditionally, Swedish school meals constituted a collective social reform in the developing welfare state and school lunches have been provided to all pupils in primary school since 1946. Thereby school meals came to symbolise universality and social equality, with everyone receiving the same food free of charge (Gullberg, 2006; Persson Osowski, Göranzon, and Fjellström, 2010). According to the Swedish Education Act, nutritious school meals must be offered to all children in compulsory school (Swedish Parliament, 2010). However, it is not further specified *what type* of diet the children are entitled to. The advisory guidelines for school meals stress that one or more cooked dishes should be served daily, ideally including a vegetarian dish (National Food Agency, 2013), and the provision of a vegetarian option served to all has become more common (Patterson and Schäfer Elinder, 2015). When it comes to religiously sanctioned food, the guidelines only state that schools should take religious needs into account as far as possible (Swedish National Food Agency, 2013). This leaves room for negotiations and discussions about what should be served and why. Moreover, in the past few years, a clear trend has emerged, namely a shift towards an increasing diversification and more calls for individual solutions, and this may put high demands on school meals, both financially and logistically (Persson Osowski, 2012). This development has taken place in line with the social force of individualisation, which Warde (1997, p. 181) describes as personal choice having a prevailing influence on people's eating habits, rather than common patterns of consumption, which were dominant in the past. At the same time, a countertendency towards individualisation has occurred through the process of communification, which entails people creating imagined communities as a result of a lack of social belonging in the increasingly individualised society (Warde, 1997). Warde (ibid) singles out religious and ethnic groups as some of those more 'communified' than others. This development raises questions about how the collective Swedish school meal may be combined with the religious preferences of minority ethnic communities. In practice, this may entail that meat must be slaughtered according to religious precepts (Enkvist, 2013) such as halal in Islam, or that milk and meat must be kept separate throughout the production chain and not served together as part of a meal, known as *kosher* in Judaism (Dugan, 1994).

To our knowledge, no previous studies have looked at online media debates with regard to religiously sanctioned school meals. However, a few studies have been published on food, religion and ethnic minorities in the school setting and home context, illustrating how religiously sanctioned food can lead to inclusion and exclusion. A Danish study (Andersen, Holm, and Baarts, 2015) discussed how the lunch-pack-versus-the-communal-school-meal arrangement influences social life. One of the conclusions was that the intention to even out social inequalities with a free school meal may be overridden by a new type of social exclusion as some children were unable to eat the communal meal for religious reasons. Religious minority children may be categorised as different from the majority children (Giovine, 2014), which may create a stigma in the meal situation, something that research on Swedish young people who need a special diet for medical reasons has shown (Olsson et al., 2009). According to Giovine (2014), who has looked at the situation of Muslim children in Italy, children may be caught in the middle between the conflicting demands of their parents, the school, the catering service, the public administration and their peers. This was exemplified in a Danish study (Karrebæk, 2014) of food and Muslim children as an ethnic minority. In Denmark, children usually bring their own food to school and, in this study, Muslim children often brought food that was different from the food items classified as healthy by the common social norm. This resulted in the children being held accountable for what their parents put in their packed lunches. A UK study

(Twiner et al., 2009) has shown that parents may see packed lunches as a 'safer' option. Parents were concerned about competing values in the school environment and doubted that the caterer had properly understood the need for religious diets. Lunch packages may, however, serve another purpose. For example, according to Nukaga (2008), food brought from home may function as a cultural object that is used to 'do ethnicity' (p. 371). This American study showed that by using food in gift-giving, sharing and trading, children both marked and muted ethnic boundaries. Although traditional eating habits are often considered important to ethnic minorities, another Danish study (Nielsen et al., 2015) showed that parents often want their children to feel at home both in the minority and majority cultures, one of the reasons being that their children would avoid stigmatisation. Research from Norway also demonstrated that ethnic minority children influence what their families eat (Mellin-Olsen and Wandel, 2005), which makes it particularly interesting to study the school meal, a meal where children encounter eating habits that may be different from home.

In sum, a few studies have covered food, religion and ethnic minorities in the school setting, but none has focused specifically on religion and food in the Swedish context, with our legally guaranteed free school meals. Religiously sanctioned school meals are already being discussed on social media as part of an ongoing societal debate about religion in Sweden, and this chapter focuses on what norms and values are being communicated in this online sphere. This is important, as the Swedish welfare state needs to relate to these opinions when providing welfare services, including school lunches, as how these conflicting views are resolved sets the scene for what should apply in public space in the Swedish welfare society. What is more, the media has always been a platform for religious messages, and in recent decades the internet has been added as another medium used for discussing religious issues. This information revolution has led to religion being talked about on millions of webpages, and information is now available to more people than ever before in history (Lövheim, 2007). Since our social world is becoming increasingly digital, research needs to move online in order to understand society (Kozinets, 2010). The present chapter aims to illuminate what is expressed on social media about religiously sanctioned school meals.

Materials and methods

Data collection

This study is based on Swedish blog and internet forum posts. The data were collected over the course of one month, starting on 23 February 2016 and ending a month later on 23 March 2016. The inclusion criteria were as follows: (1) only Swedish blogs and forums with entries written in Swedish, (2) for research ethics reasons, only public webpages requiring no login and (3) only entries posted in the 2000s. Due to the large number of blogs and internet forums available, it was necessary to make a selection. In qualitative studies, it is common to use purposive sampling (Patton, 2002). However, since the present study was exploratory, a standardised search procedure, with predetermined search terms, was used instead. These search terms were religion + skolmat + blogg (religion + school meals + blogs, in Swedish), producing 21,400 results, and religion + skolmat + forum (religion + school meals + internet forums, in Swedish), producing 13,600,000 results. To narrow down these numbers, we included all relevant blogs and internet forums that appeared on the first five pages for each search. The original posts and the comments on them were copied into two separate Word files, one for the blogs and one for the forums, respectively. During this process, we anonymised all online pseudonyms, names and other information that could be used to identify the blogs and forums and excluded posts irrelevant to the aim of the study. This resulted in the volume of text equivalent to over 300 pages of plain text in Word. Although we had planned to use more search terms, the data collection ended at this stage as it was judged impossible to qualitatively analyse more material. The collected material comprised 12 blogs and 7 internet forums, which were heterogeneous in their characteristics. The posters' background characteristics were not always specifically stated, but both child and adult posters were represented and the blogs and internet forums had varying profiles, ranging from religion to food and health and from gaming to animal protection. Some of them took on a more political character, positioning themselves as either antiracist or for the preservation of Swedish culture and taking a critical stance towards immigration and immigrants.

Analysis

The two Word documents were imported into the qualitative data analysis software NVivo Plus, version 11 (QSR International). We conducted the analysis using an inductive thematic analysis (Braun and Clarke, 2006) within the social constructionist paradigm. According to social constructionism, how we perceive the world results from the historical and cultural context that we live in and the common-sense knowledge constructed about our social reality is sustained by social processes (Burr, 2003). In these modern times, the internet cannot be disconnected from society as a whole (Lövheim, 2007) and although this study only focuses on the social context online, we assume that this reality is both influenced by and will influence the rest of our social reality. The analysis started with a reading and rereading of the data, and then noting initial ideas (Braun and Clarke, 2006). The coding was initially kept close to data. During this process, we identified two preliminary themes, which were thoroughly reviewed and refined during the coding process. This inductive process involved the constant creation and recreation of subthemes and codes. As a last step, a thematic map (Braun and Clarke, 2006) was produced, resulting in a final framework presented in the results section. Both authors were involved in the analysis process, including deciding on the final framework and themes and in verifying the results.

Ethical considerations

The present study was based on already available data (Silverman, 2006). This is positive from a research-ethical point of view as no new material had to be collected and no new study participants had to be involved. However, the internet is a relatively new source of research, and therefore it is vital to make informed ethical decisions (Markham and Buchanan, 2012; Simunaniemi, 2011). According to Elliott, Squire and O'Connell (2017), referring to the Association of Internet Researchers' recommendations, it is not possible to specify in advance when research might cause harm, as internet research is heterogeneous and dynamic. As a result of this, a case by case approach is needed. A few ethical guidelines are available and according to Bruckman (2002), researchers may, without consent, freely analyse and quote information found on the internet as long as it is publicly archived, no password is needed for accessing the information, no site prohibits it, or the topic is not sensitive. As our study addresses religion, the topic may be considered sensitive. For this reason, an application for the ethical vetting of research involving humans was sent to the ethical review board in Uppsala (Ethical Review Boards, 2016). The board judged that the study did not require ethical vetting pursuant to the Act on Ethical Review of Research Involving Humans. Therefore, only an advisory statement, including no objections to the research project, was given instead (reference number 2015/471).

According to Walther (2002), anyone who publicly communicates on the internet cannot expect what they write to be private. However, posters do not always expect that their entries may be used for research purposes but, due to their anonymity and going under pseudonyms, it is difficult, or even impossible, to ask for informed consent. Kozinets (2010) maintains that you only need consent if the research involves interaction or an intervention online, which was not the case in our study. Nevertheless, the more vulnerable a poster is, the greater the researcher's responsibility to protect that individual (Markham and Buchanan, 2012). Therefore, our study collected no online pseudonyms or names and no blog or forum names were presented as they could be traced to the poster (Bruckman, 2002; Kozinets, 2010). Verbatim quotes, however, were used since they are an important qualitative aspect of qualitative research, allowing the reader to assess the trustworthiness of a study (Guba and Lincoln, 1994). These quotes were translated into English, which makes it more difficult, but not impossible, to find the original source. Due to this, we gave careful consideration to what should or should not be published.

Results

The two main themes, entitled *Traditional school meals* and *Cultural relativist school meals*, are presented with illustrative quotes and are summarised in the framework in Figure 4.1. The framework illustrates how the ideas expressed on social media fall along two continuums ranging from statements on secularism to freedom of

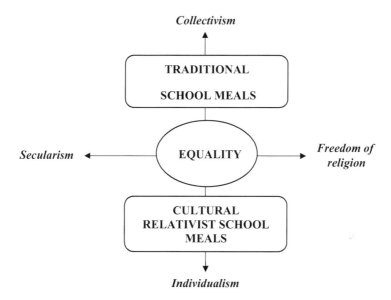

Figure 4.1 A framework depicting the discussions about religiously sanctioned food on social media.

religion and from collectivism to individualism. The posts on social media took on different positions along these two continuums, ranging from more neutral in the middle to the extreme positions at the end for both continuums. The first continuum comprised thoughts ranging from a secular society, school and school food to food as an expression of freedom of religion. The second continuum covered different opinions on whether school meals should be collective or to what degree individual solutions should be provided. In spite of the different opinions put forward in the social media posts, a central finding was a striving for equality in the school meal situation. In the traditional theme, we found what we chose to call an *equal-collectivistic* standpoint, where the arguments gave prominence to the importance of serving the same food to everyone, whereas the cultural relativist standpoint insisted on an *equal-individualistic* perspective, allowing people food according to their needs and wishes.

Traditional school meals

The posts categorised under *Traditional school meals* stemmed from the view that foodways which are traditional for the Swedish school meal should be the norm. This entailed collective meals consisting of the same food for everyone, which is equal for all, thereby representing the *equal-collectivistic* standpoint. This theme could be seen as ethnocentric in many ways. Of central importance was that

Swedish culture and traditions should be preserved by including traditional food and dishes. The advocates of this position repeatedly gave prominence to a secular society in general and a secular school in particular. This view entailed that there should be no religious elements in public schools and, as a result of this, religiously sanctioned food should generally not be served. The rationale behind this was that society should not be governed by religion and that school is to be secular in accordance with Swedish law. Criticism was especially directed at what was referred to as the 'Islamisation' of school and society. This was illustrated by some of the writers attacking some schools for having opted to serve soya or halal meat to everyone and less or no pork in order to meet the rising number of religious requests. Keeping pork on the menu thus became a symbol of no religious conformity, as exemplified in the following internet forum quote, which discussed the issue of no longer serving pork:

> It's simply insane that a school in Sweden would drop something like pork, which is rooted in the country, just because non-Nordic immigrants – with a religion that traditionally is non-existent in Northern Europe – don't eat it.

The repudiation of religious food was clearly posited in many posts and in some as an expression of patently xenophobic views. Others voiced their opinions based on the need for assimilation; they wrote that religious people should not have the right to make demands and that immigrants should conform to Swedish society and not the other way round, as exemplified in the following internet forum quote:

> They get food, but do they want to eat it? Nah, why? They believe in and live their lives according to fairy tales. Why should a non-Muslim have to conform to a Muslim, why isn't it the other way round? When in Rome, do as the Romans do, isn't that right. . . ?

Freedom of religion was mentioned in a reversed sense, i.e. school meals should be secular, and non-religious people or people of a different religion should not have to eat food that another religion has sanctioned, with halal food often being cited. On the other hand, some posters did say that there are people who, for religious reasons, need a special diet and in this regard freedom of religion was also visible in the material. However, it was made clear that it is not the school's responsibility to provide the pupils with religiously sanctioned food, but it is up to the individuals themselves how this should be solved. This clearly exemplified the equal-collectivistic standpoint, where the same rules apply to everyone and where everyone is served the same food. The most commonly mentioned solutions were that these pupils should provide their own meals, for instance by bringing their own food to school, or that they should have the vegetarian option when the meat served is not religiously approved. Other solutions cited were that pupils needing religiously sanctioned food could attend a religious independent school or be homeschooled, as exemplified in the following comment on a blog entry:

> Homeschool your children yourself instead of making more and more absurd demands on the school, the school environment, school food, etc.

Cultural relativist school meals

Cultural relativist school meals were the opposite of their traditional counterparts; i.e. beliefs and individual school meal requests should be seen from the individual's perspective, with differences and diverse foodways being accepted and acknowledged. These posts thus represented the *equal-individualistic* standpoint, stressing personal choice as important when striving for equality. Both secular and religious values were expressed and the posts that acknowledged freedom of religion implied setting one's own cultural values aside and accepting differences. These posts supported tolerance and conforming to the new Swedish society as it looks today, as illustrated in one of the internet forums:

> It's not about society conforming to Muslims! It is about society conforming to the needs of its citizens. A modern Sweden must be a society that is tolerant and respects minorities so that no one has to be subjected to discrimination or degrading treatment.

The acknowledgment of the right to religiously sanctioned food in a secular society, the expression of inclusion and the right to different beliefs opposed to xenophobia were among the views voiced by the cultural relativist posts. This was done by advocating the provision of religiously sanctioned food for the individual needing it, thereby implying that for it to be an equal meal, the food served must satisfy everyone, irrespective of the reason being religious, ethical or medical. This was exemplified in the following blog quote, which also underpinned that religiously sanctioned food is compatible with a secular school because it is related to freedom of religion:

> However, there is halal food at school, but it is categorised as a special diet served to pupils who have chosen it. Then it is an active choice, which is not part of the norm, but something that the believers themselves have to take responsibility for. This is, in other words, part of freedom of religion (that there are alternatives for those of another religion) and not part of a determination to incorporate Swedish schools into an Islamic world (which some seem to believe, or have had others believe). Just like there are other special diet options for those pupils who don't eat meat at all or are allergic to certain ingredients.

Posts that expressed modern food ideologies as an individual need comparable to religiously sanctioned food were also categorised under the cultural relativist theme. This transformation and comparison concerns predominately vegetarian and vegan diets; environmentally sustainable food; a gluten-free diet; a low-carbohydrate, high-fat diet (LCHF) as well as only Swedish meat for ethical reasons. The comments included different arguments for why a specific diet was important in the school meal setting. Taking vegetarian food as an example, these discussions were closely intertwined with environmentally sustainable food as a meatless diet was put forward as good for the environment. Although promoting a vegetarian option at school was conveyed using other reasons such as health, the ideological grounds for protecting the environment were particularly prominent. The advocates of the various food ideologies regarded these régimes as diets that one should also be entitled to receive at school. Moreover, its spokespeople sometimes portrayed themselves as on a mission to gain acceptance for their food claims. The food ideologies that made health claims about their diets, like the spokespeople for a gluten-free diet and LCHF, also directed their criticism at school food, which they described as unhealthy and as a product of the authorities' established definition of what constitutes a healthy diet. In relation to other special diets served, the advocates also wanted their food ideologies to be acknowledged as equally valid, as illustrated in the following example from one of the internet forums:

Person 1:   'Like people with vegetarian or religious convictions, I want the right to, without a medical certificate, have my conviction about what my children should avoid eating respected'.
Person 2:   'Maybe it is time to start the religious community's "LCHF mission . . ."'

This position, moreover, entailed accommodating so that there is food at school which everyone can accept. The solutions for achieving this differed, however. There was disagreement on how many dishes should be served at school each day, but several dishes to choose from was seen as an option so that everyone can find something to eat. Others favoured tailoring the main dish to suit as many individuals as possible. One example of this was to serve less or no pork, and replace it with, for instance, chicken. Others advocated treating religiously sanctioned food as a special diet for the individual. Although the solutions mentioned differed, these statements were all clear representations of the equal-individualistic standpoint.

Discussion

The comments in different posts on Swedish internet blogs and forums constructed the school meal as a point of tension between views characterised as traditional or cultural relativistic. However, the phenomenon included different standpoints, falling along two continuums, ranging from statements on secularism to freedom of religion as well as from collectivism to individualism. In essence, the social media posts represented opinions on what the school meal should be like in relation to social inclusion and exclusion and therefore the concept of commensality, i.e. sharing food together, becomes applicable. Sobal (2000) has described how people who belong to different social networks are thereby also part of different commensal circles, and those circles define 'us' versus 'them'. In other words, for those involved, commensality integrates, whereas it disintegrates for those who are not (Fischler, 2015), leading to either inclusion or exclusion. In this respect, the social media posts about religiously sanctioned school meals represented opinions on the changing Swedish society in general and became a symbol of integration and disintegration regarding the school meal situation in particular. According to Fischler (2015), commensality has undergone changes in the West owing to the process of individualisation. As a result of this, he distinguishes between two dimensions of commensality, namely, a more traditional *communal dimension*, with the people having a meal together forming a communion and an individualised *contractual dimension*, with people spending time together around a meal, but where what is eaten is a product of negotiation.

Traditional school meals

Traditional school meals were a clear representation of the communal dimension, which symbolises a collective meal where the same food is eaten, common values are shared and it is not socially accepted to demand that alternative foods be served (Fischler, 2015). In the context of Swedish school meals, this dimension entailed ethnocentric values, with a traditional Swedish menu shared by everyone and where secular ideals are favoured. Commensality, moreover, involves cultural codes about eating patterns and different types of food stand out as being better than others (Sobal, 2000). From the traditional point of view, favouring traditional Swedish food items like pork became a means of preserving Swedish secular values and thus opposing the influence of religion and the food items representing this, such as halal meat. Here the concept gastronationalism becomes applicable. According to DeSoucey (2012), gastronationalism describes nationalist sentiments, national attachment and how food forms an integral part of national identity, and the traditional standpoint was a clear representation of this. The traditional position also involved statements about assimilation and the need for immigrants to conform to Swedish society, and as such the school meal became a symbol for demarcating what may be accepted or not. On the other hand, this theme also comprised clearly xenophobic statements. Mennel, Murcott and van Otterloo (1992, p. 117) wrote that commensality, by its inclusiveness, also produces xenophobic norms and exclusion and the traditional position showed examples of this. With the solutions mentioned, like bringing your own food to school or being taught at home or at a religious independent school, you naturally become excluded from the school meal or even Swedish society as a whole. Sweden is a secular country, but it also has a long tradition of Protestantism, and according to Tange Kristensen et al. (2012), this religion is the only one with no food-related rituals and rules. The conception that the communal

Swedish school meal is secular can thereby be seen as a blind spot where the absence of religion makes people less understanding of religious aspects of food and meals. Fischler (2015) states, however, that Protestant countries are more prone to accept personal dietary requirements and thereby a contractual dimension of commensality.

Cultural relativist school meals

What Fischler (2015) has described as the contractual dimension was represented in the cultural relativist position. As previously mentioned, the contractual dimension of commensality entails a negotiation between those sharing a meal based on personal preferences about what is to be eaten. In the Swedish school meal situation, with several dishes being served, those at the same table do not necessarily share the same food, thereby taking the contractual dimension even further. The cultural relativist position left room for individual differences and preferences, thus representing the process of individualisation that has occurred in the western world (Fischler, 2015; Warde, 1997). Although this position accepted that individual demands should be acknowledged with regard to freedom of religion, there was disagreement on how many dishes are to be served, with some preferring that the main dish should be tailored instead to suit as many as possible. Others suggested that a special diet should only be provided for those who need religiously sanctioned food but, either way, religious foodways were to be recognised.

The cultural relativist position also comprised discussions on secular diets. In some, but not all cases, the ideological foodways described as, for example, an LCHF or a gluten-free diet, could be characterised as quasi-religious because they showed elements of a traditional religion, but under secular circumstances (Zeller, 2015). There is no stipulated definition of what religion means, not even in a legal sense (Enkvist and Nilsson, 2016, p. 93) and thereby it is preferable to use the term 'quasi-religion' to describe foodways like veganism, gluten-free diets or diets like Paleo when embraced by cultural movements. However, these foodways bear many similarities to religions in the more traditional sense in that they create a sense of identity, community and meaning for the individuals following these diets. For instance, veganism often entails more than special eating practices but also codes of ethics and morality, life-guiding texts, communities of practitioners and an identity; however, at the same time, it contrasts with traditional religions in that it appeals mainly to the most secular. Also contributing to the popularity of these fad diets are the celebrities who proclaim they adhere to them (Zeller, 2015). It is important to stress that simply eating a special diet does not automatically mean that the foodways are quasi-religious. But there are similarities to religion, and this also constitutes an example of the process of communification, e.g. what Warde (1997) described as people's creations of imagined communities in the absence of social belonging in an individualised society.

The solutions mentioned by the cultural relativist camp may come across as mainly inclusive, but they are not without problems. For instance, religiously sanctioned meals could create social exclusion and stigmatisation if not everyone can eat the common meal (Andersen, Holm, and Baarts, 2015; Giovine, 2014). Moreover, multiple personal dietary requests challenge the limited economic resources of the school lunch, and the degree to which religious principles should be acknowledged and how these values are fused with Swedish secular traditions are not uncomplicated matters. Different solutions tried in the United Kingdom and mentioned in the analysed social media, such as eliminating staple meats like pork, serving halal meat to everyone, or providing vegetarian options may work in some schools, but may be unacceptable in others (Twiner et al., 2009). Ideally, the school meal could become what Anderson (2011) refers to as a cosmopolitan canopy, which forms a pluralistic space where racially, ethnically and socially diverse people come together and are informally educated about each other's differences, something that Anderson (2011, p. xvii) calls 'folk ethnography'. This may change the common-sense understandings that people have of one another and lead to tensions being defused, a civilised coexistence and people being less ethnocentric (Anderson, 2011).

Equality

Fischler (2015) linked the two dimensions of commensality to different cultural contexts, with the communal dimension being found in continental countries like France and the contractual dimension in the more individualised Anglo-Saxon countries. Thus, there seems to be a Swedish paradox in that we detected both dimensions in the same cultural context. The answer to this paradox may lie in the characterisations of cultural values that distinguish Sweden from other parts of the world. As mentioned in the introduction, in the World Values Survey, Sweden found itself at the very extreme position, both for secular-rational values and self-expression values (Institute for Comparative Survey Research, 2017). To explain the two dimensions of commensality within the same cultural context, we would like to elaborate on the value of equality, as a representative of self-expression and as a traditional value of the Swedish welfare state. This is also true for the school meal as a welfare service, which is embedded with values such as universal benefits and social equality (Gullberg, 2006; Persson Osowski, Göranzon and Fjellström, 2010). Irrespective of traditional opinions or the opposite cultural relativist position, a striving for equality could be identified in the arguments of both themes. In the traditional theme, we found what we chose to call an *equal-collectivistic* standpoint, where the arguments gave prominence to the importance of serving the same food to everyone. This position reflected the more traditional values of the Swedish school meal as being a part of the collective Swedish welfare state (Persson Osowski et al., 2010). On the other hand, representatives of the cultural relativist perspective insisted on an *equal-individualistic* perspective, allowing people food according to their needs and wishes.

This position stressed personal choice and was clearly influenced by the process of individualisation. Equality, one of the most important aspects of the Swedish school meal, seemed to thereby link diametrically opposed opinions together. Maybe the striving for equality found in both traditional and cultural relativist standpoints can be referred to as 'commensal equality'. This notion is based on the idea that both sides made great efforts to produce an equal meal for everyone, one by serving the same food to everyone and the other by serving food that will satisfy everyone.

Concluding remarks

Children's eating is both the public and the private domain's responsibility, and as a result of this school meals find themselves in the borderland between family and state (Gustafsson, 2002). In Swedish schools – a public space – a communal lunch is shared. Still, increasing demands have been made that school meals are to assume the same values as a domestic meal (Persson Osowski, 2012) and that meals are personalised, which is in line with the social force of individualisation (Warde, 1997). It is in this context that the online discussions about religiously sanctioned school meals began, and at a time when migration is increasing it becomes even more important to illustrate what is expressed on social media regarding what should apply in a public space. How this issue will be settled sets the scene for how conflicting values and norms are to be resolved, not only in this particular situation, but it may also affect how conflicts within the transforming welfare state are resolved in general. Thus, the question raised is not only whether Swedish school meals are to be traditional or cultural relativist but also in the broader, bigger picture, whether Swedish society is to conform to the individual or vice versa.

Methodological considerations

The present study has some limitations. As this is a first exploratory qualitative study based on a limited number of blogs and internet forums, the results are not transferable to all forms of social media in Sweden. Moreover, as many writers keep their identities and personal characteristics anonymous on the internet, it is difficult to describe in detail who has been included in the material or describe the context in which the discussions took place. However, it is likely that some groups may be underrepresented in the material, for instance ethnic minorities. Moreover, blogs and posts on internet forums are an asynchronous form of interaction and thereby differ from synchronous face-to-face interaction, for instance, in that it involves fewer social cues (Schiek and Ullrich, 2017). This may lead to a less nuanced language and insensitive tone being used. Using a small number of search terms naturally also limited what was captured on the social media analysed. Despite these limitations this study is, to our knowledge, the first to use naturally occurring data (Silverman, 2006) in the form of social media discussing this topic, thereby allowing insight into norms and values regarding religiously sanctioned school meals in a multicultural country with a culturally and religiously diverse society.

Conclusion

What was expressed on social media depicted religiously sanctioned school meals as either traditional, advocating a collective, secular school meal where religiously sanctioned food is not provided for, or cultural relativist, acknowledging individual differences and where religiously sanctioned food is catered for. Religiously sanctioned school meals are just one example among many issues that the Swedish welfare state has to solve in a society that has become increasingly diverse. How the welfare state manages to bring together the traditional and cultural relativist standpoints may lead to either integration or tension in society as a whole and in societal institutions such as schools.

References

Andersen, S.S., Holm, L. and Baarts, C. (2015). School meal sociality or lunch pack individualism? Using an intervention study to compare the social impacts of school meals and packed lunches from home. *Social Science Information*, 54(3), pp. 394–416. doi:10.1177/0539018415584697.

Anderson, E. (2011). *The cosmopolitan canopy: Race and civility in everyday life*. Vol. 1. New York: W.W. Norton & Co.

Braun, V. and Clarke, V. (2006). Using thematic analysis in psychology. *Qualitative Research in Psychology*, 3(2), pp. 77–101. doi:10.1191/1478088706qp063oa.

Bruckman, A. (2002). *Ethical guidelines for research online*. Available at: www.cc.gatech.edu/~asb/ethics/

Burr, V. (2003). *Social constructionism*. 2nd ed. London: Routledge.

DeSoucey, M. 2012. Gastronationalism. The Wiley-Blackwell Encyclopedia of Globalization: John Wiley & Sons, Ltd.

Dugan, B. (1994). Religion and food service. *The Cornell Hotel and Restaurant Administration Quarterly*, 35(6), pp. 80–85. doi:http://dx.doi.org/10.1016/0010-8804(95)91833-2.

Elliott, H., Squire, C. and O´Connell, R. (2017). Narratives of normativity and permissible transgression: Mother´s blogs about mothering, family and food in resource-constrained times. *Forum: Qualitative Social Research*, 18(1). http://dx.doi.org/10.17169/fqs-18.1.2775

Enkvist, V. (2013). *Religionsfrihetens rättsliga ramar*. PhD diss. Uppsala University.

Enkvist, V. and Nilsson, P.-E. (2016). The hidden return of religion. Problematising religion in law and law in religion in the Swedish regulation of faith communities. In: A.-S. Lind, M. Lövheim and U. Zackariasson, eds., *Reconsidering religion, law and democracy. New challenges for society and research*. Lund: Nordic Academic Press, pp. 93–108.

Ethical Review Boards. (2016). *Welcome to ethical vetting*. Available at: www.epn.se/en/start/

Fischler, C. (2015). Introduction. In: C. Fischler, ed., *Selective eating: The rise, meaning and sense of personal dietary requirements*. Paris: Odile Jacob, pp. 15–35.

Giovine, R. (2014). Big demand(s), small supply – muslim children in Italian school canteens: A cultural perspective. *Young Consumers*, 15(1), pp. 37–46. doi:10.1108/YC-03-2013-00359.

Guba, E.G. and Lincoln, Y.S. (1994). Competing paradigms in qualitative research. In: N.K. Denzin and Y.S. Lincoln, eds., *Handbook of qualitative research*. Thousand Oaks, CA: Sage, pp. 105–117.

Gullberg, E. (2006). Food for future citizens: School meal culture in Sweden. *Food, Culture and Society*, 9(3), pp. 337–343. doi:10.2752/155280106778813279.

Gustafsson, U. (2002). School meals policy: The problem with governing children. *Social Policy and Administration*, 36(6), pp. 685–697. https://doi.org/10.1111/1467-9515.00311

Institute for Comparative Survey Research (2017). *World values survey*. Available at: www.worldvaluessurvey.org/wvs.jsp

Karrebæk, M.S. (2014). Rye bread and halal: Enregisterment of food practices in the primary classroom. *Language & Communication*, 34, pp. 17–34. http://dx.doi.org/10.1016/j.langcom.2013.08.002

Kozinets, R.V. (2010). *Netnography. Doing ethnographic research online*. London: Sage.

Lövheim, M. (2007). *Sökare i cyberspace: ungdomar och religion i ett modernt mediesamhälle*. Stockholm: Cordia.

Markham, A. and Buchanan, E. (2012). *Ethical decision-making and internet research. Recommendations from the AoIR ethics working committee version 2.0*. Available at: http://aoir.org/reports/ethics2.pdf

Mellin-Olsen, T. and Wandel, E. (2005). Changes in food habits among Pakistani immigrant women in Oslo, Norway. *Ethnicity & Health*, 10(4), pp. 311–339. doi:10.1080/13557850500145238.

Mennel, S., Murcott, A. and van Otterloo, A.H. (1992). *The sociology of food: Eating, diet and culture*. London: Sage Publications Ltd.

National Food Agency (2013). *Good school meals. Guidelines for primary schools, secondary schools and youth recreation centres*. Available at: www.slv.se/upload/dokument/mat/mat_skola/Good_school_meals.pdf

Nielsen, A., Krasnik, A. and Holm, L. (2015). Ethnicity and children's diets: The practices and perceptions of mothers in two minority ethnic groups in Denmark. *Maternal & Child Nutrition*, 11(4), pp. 948–961. doi:10.1111/mcn.12043.

Nukaga, M. (2008). The underlife of kids' school lunchtime. Negotiating ethnic boundaries and identity in food exchange. *Journal of Contemporary Ethnography*, 37(3), pp. 342–380. doi:10.1177/0891241607309770.

Olsson, C., Lyon, P., Hörnell, A., et al. (2009). Food that makes you different: The stigma experienced by adolescents with celiac disease. *Qualitative Health Research*, 19(7), pp. 976–984. doi:10.1177/1049732309338722.

Patterson, E. and Schäfer Elinder, L. (2015). Improvements in school meal quality in Sweden after the introduction of new legislation – a 2-year follow-up. *European Journal of Public Health*, 4(1), pp. 655–660. https://doi-org.ezproxy.its.uu.se/10.1093/eurpub/cku184

Patton, M.Q. (2002). *Qualitative research & evaluation methods*. 3rd ed. London: SAGE.

Persson Osowski, C. (2012). *The Swedish school meal as a public meal. Collective thinking, actions and meal patterns*. PhD diss. Uppsala university.

Persson Osowski, C., Göranzon, H. and Fjellström, C. (2010). Perceptions and memories of the free school meal in Sweden. *Food, Culture and Society*, 13(4), pp. 555–572. doi:10.2752/175174410x12777254289420.

QSR International. *NVivo plus* (Version 11).

Schiek, D. and Ullrich, C.G.(2017). Using asynchronous written online communications for qualitative inquiries: A research note. *Qualitative Research*, 17(5), pp. 589–597.

Silverman, D. (2006). *Interpreting qualitative data: Methods for analyzing talk, text and interaction.* 3rd ed. London: SAGE.

Simunaniemi, A.-M. (2011). *Consuming and communicating fruit and vegetables: A nationwide food survey and analysis of blogs among Swedish adults.* PhD diss. Uppsala university.

Sobal, J. (2000). Sociability and meals: Facilitation, commensality and interaction. In: H.L. Meiselman, ed., *Dimensions of the meal: The science, culture, business, and art of eating.* Gaithersburg: Aspen Publishers Inc, pp. 119–133.

Swedish Parliament (2010). *Skollag (2010:800).* Available at: www.riksdagen.se/sv/Dokument-Lagar/Lagar/Svenskforfattningssamling/Skollag-2010800_sfs-2010-800/?bet=2010:800#K10

Tange Kristensen, S. and Houlind Rasmussen, U. (2012). Når mad er gud – Moderne sundhetsdyrkelse. In: L. Holm and S. Tange Kristensen eds., *Mad, mennsker & måltider.* Copenhagen: Munksgaard, pp. 425–438.

Twiner, A., Cook, G. and Gillen, J. (2009). Overlooked issues of religious identity in the school dinners debate. *Cambridge Journal of Education*, 39(4), pp. 473–488. doi: 10.1080/03057640903352457.

Walther, J.B. (2002). Research ethics in Internet-enabled research: Human subjects issues and methodological myopia. *Ethics and Information Technology*, 4(3), pp. 205–216. doi: 10.1023/a:1021368426115.

Warde, A. (1997). *Consumption, food, and taste: Culinary antinomies and commodity culture.* London: Sage.

Zeller, B.E. (2015). Totem and taboo in the grocery store: Quasi-religious foodways in North America. *Scripta Instituti Donneriani Aboensis*, 26, pp. 11–31.

Part II

Changes and challenges

Chapter 5

'I wouldn't delve into it too much'

Public concerns (or not) about the contemporary UK food supply system

Alizon Draper,[1] Val Gill, Hayley Lepps,
Caireen Roberts and Judith Green

Introduction

In the face of climate change and recent socio-political events, such as Brexit, there is renewed academic and policy interest in food systems regarding how they are structured, controlled and regulated, and how to manage ongoing and emergent risks (see e.g. Food and Agriculture Organization, 2014; Foresight, 2011; Food Standards Agency, 2016; IPES-Food, 2017). Globalisation has meant that food supply chains (FSCs) have become increasing complex with many foods and ingredients travelling long distances and passing through many steps from harvesting, processing, packaging and distribution before they enter the retail arena and are available to us as consumers to purchase, prepare and eat. It has been speculated these changes in FSCs have led to new risks and new anxieties on the part of the public despite the paradox that our food supply (for now) is more plentiful and safe than ever before (Beardsworth and Keil, 1997; Caplan, 2000; Jackson, 2015). It has also been argued that the increased complexity of the FSC and the consequent distancing of most people from most steps in FSCs has created a sense of alienation, which has in turn fostered new food fears and anxieties that focus on the safety rather than the availability of food (Gofton, 1990). Further the food scares of the 1990s onward have been seen as creating new sites of social anxiety (Unger, 2003). Small-scale qualitative studies do confirm that some people do feel a sense of 'disconnection' with the food system and actively seek to engage with small-scale and local food networks, such as farmers' markets and box schemes (Kneafsey et al., 2008). But, despite the increasing policy concern about the state of national and international food systems and supply chains, there is surprising little empirical evidence on people's understandings and concerns regarding these (Darnton, 2016).

Theories of modernity and the risk society (Giddens, 1991, Beck, 1992) posited that attunement to risk and anxiety are endemic to contemporary life. The 'risk society' thesis was not that modern societies faced more risks than previous ones, but that future-orientation and reflexivity shaped a preoccupation with risk, and a privatisation of risk, such that individual citizens (or food consumers) were obliged to be constantly attuned to risk, and responsibilised to manage and maximise their own health and security: 'Doubt, a pervasive feature of modern critical reason, permeates into everyday life . . . and forms a general existential dimension of the contemporary social world' (Giddens, 1991, p. 3).

In this context, to be modern is to be engaged in what Giddens called an 'ever present exercise' (Giddens, 1991, p. 124) of risk assessment, in which we are constantly balancing the probabilities of particular harms and benefits of actions and decisions for us as individuals, into uncertain futures. Of course, this does not reduce the population to constant anxiety over what to do. Rather, a pervasive trust in the continuing, normal, everyday nature of the world enables most people, most of the time, to continue making routine and habitual decisions about what to do: how to select food to eat from the cosmopolitan range on offer; how to prepare it; how to avoid potential risks.

Contemporary food supply systems typify some of the challenges of knowledge in this conception of modernity. Given the fine divisions of labour in modern societies, and the complex, dispersed organisational forms of globalised food supply systems, the 'system' is knowable in theory but, in practice, impossible for any individual to know fully. Understanding all the technical, legal and social components of those long chains which link, say, agricultural production to a foodstuff on a supermarket shelf, requires specialist knowledge across a vast range of arenas. Weber's essay 'Science as a Vocation'([1917) touched on this inherent tension in the status of knowledge in the modern world: that with the growth in legitimacy for rational explanation, all things were potentially understandable, yet the field of things actually known about diminished. That which could not (even in principle) be known rationally was relegated to a denigrated realm: of superstition, faith, religion, as the post-Enlightenment world became, in Weber's phrase, 'disenchanted'. FSCs, as part of those complex food supply systems, are entangled, globalised and dispersed: but they are, in principle, organised around rational scientific and economic principles, if impossible for any individual to fully comprehend.

However, there are perhaps increasing obligations for the public to be knowledgeable about these systems, as both rational consumers and engaged citizens (Draper and Green, 2002). Many current policy initiatives around food in the UK and elsewhere focus on the provision of information to the public with the expectation that this information will be acted upon by consumers. Packaged foods in industrial countries are now freighted with labels of different kinds conveying information about not only ingredients but also nutrients, allergens, guidance on use and storage, through to country of origin. There is also a growing emphasis on 'empowering' and engaging the public (see e.g. Food Standards Agency, 2016) as well as a growing movement in the civil society sector for consumers to become 'food citizens', a broader configuration in which people are expected to 'think big' (New Citizenship Project, 2017).

In the context of potential breaches in the routine trust that individuals have in the food supply system, typified by 'food scares', there are urgent questions about the ways in which knowledge about that system is organised: what does the public know, what can it know and what are the implications of that knowledge? To explore these questions, this chapter draws on data from a project funded by the Economic and Social Research Council and the Food Standards Agency as part of the Global Food Security programme Understanding the Challenges of the Food System. This was established in 2011 to bring together policy-makers and researchers in the UK to develop responses to the challenges posed by biological, environmental and socioeconomic shocks in a changing world. The study reported here, Public Perceptions of the UK Food System: Public Understanding and Engagements, and the impact of crises and scares, comprised three strands of empirical data collection: a social media analysis of Twitter during the UK horsemeat crisis in 2013; a series of deliberative workshops; and a quantitative survey of public concerns. This chapter presents the findings from the deliberative workshops. The aims of these were to map public understandings of, and engagement with, FSCs and how these might vary across different population groups and regionally, and to explore the nature of peoples' trust in different actors involved in FSCs, in order to understand how any future scares or scares related to FSCs could best be managed. The study was led by the National Centre for Social Research (NatCen).

Design and methods[2]

Data collection

The data are drawn from 24 focus groups (N = 240 total), held in six deliberative workshops that were conducted January 2015 to February 2015 in locations across the UK. The workshops were designed to explore people's understandings and concern regarding FSCs, trust and responsibility in both a 'normal state' and during 'food scares' (data on the latter are not reported here). The workshop design was adopted because it offered the opportunity to capture data from a wide range of participants in a concentrated period. Deliberative methods are a participatory approach that advocates actively involving 'the public' in decision-making processes and involve using stimulus material to enable the research to go beyond participants' surface knowledge and understanding and to develop solutions or strategies around policy issues (de Koning and Martin, 1996). They were used here because they offer a discursive and dynamic context in which to consider complex issues. This is particularly appropriate for a topic such as FSCs that may not be a 'front of mind' issue for most people. Deliberative workshops also provide a context in which people can explicate their concerns and decision making.

The workshops lasted approximately three hours and followed a uniform structure (Figure 5.1). They commenced with introductory plenary sessions involving all participants to explain the purpose of the study and the format. Participants were then split into four groups, involving approximately ten participants per

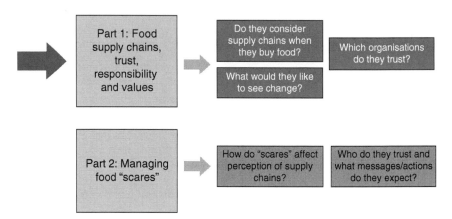

Figure 5.1 Workshop format.

group to discuss firstly participants' understandings of the FSC in a 'normal state' and secondly their reactions to a real food scare (the 2013 horsemeat crisis) and a hypothetical scare (not reported here). Each set of group discussions was followed by a brief plenary to draw issues together from across groups.

The group discussions were facilitated using a topic guide and employed a range of deliberative and mapping techniques to encourage discussion and reflection, including small group work and feedback to the wider group, flip chart work with the whole group and reflections and discussion on the points raised, discussions of newspaper headlines from the horse meat crisis and vignettes of a hypothetical food scare.

Sampling and recruitment

Maximum variation purposive sampling was used to select six sites in the UK that represented the different administrative regions to which following devolution food safety and standards powers have been delegated (Scotland, Wales and Northern Ireland). As well as differing regulations contexts, each has had different experiences of food safety incidences in the past. The six sites selected in these regions were Plymouth, Cardiff, Bradford, Glasgow, Belfast and London. Within each location the venues for data collection were chosen to be in central locations with accessible transport links.

Within each study site, maximum variation purposive sampling was also used to recruit as diverse a sample as possible and so capture the range of public views. The primary selection criteria were age, gender and social grade (assigned by occupation using the National Readership Survey classification).[3] Ethnicity was a secondary sampling criterion that was monitored during recruitment. In London non-meat eaters, Twitter users (this was for the social media strand of the study), and parents of young children were also purposively recruited. Table 5.1 shows the profile of

Table 5.1 Workshop participants

Location	Groups			
	1	2	3	4
London	10 people aged 60+	10 Twitter users	10 people living with a child under 13 (ideally under 5)	Diet – 10 non-meat eaters
Plymouth, Cardiff, Belfast, Bradford, Glasgow	10 people aged 18 to 24; any social grade	10 people aged 25 to 59; NRS social grade ABC1	10 people aged 25 to 59; NRS social grade C2DE	10 people aged 60+; any social grade

participants across these criteria. Recruitment was carried out by a specialist recruitment agency and took place on the street using recruitment materials produced by the research team to screen potential participants for eligibility to participate.

Analysis

The group discussions were digitally audio recorded and systematically analysed using Framework analysis, a thematic approach to analysing qualitative data developed by NatCen (Ritchie et al., 2013). Following this approach thematic matrices were developed through familiarisation with the data and identification of emerging issues. Each thematic matrix represented one key theme (e.g. the key barriers to engaging with the FSC), with the column headings in each matrix relating to key sub-topics, and the rows to each focus group.

Findings

What do people know about FSCs?

Tenerife or somewhere.
Glasgow group 4 60+y
All social grades

The first set of group discussions explored participants' understanding of FSCs, their considerations when purchasing food, prompts to consider FSCs and suggestions to improve public understanding of them. To start the discussions, people were asked to map out the journey by which particular foods (banana, turkey dinosaur and prepackaged cod fillet) travelled to the UK to be available to buy (see Figure 5.2). While people could generally provide a high level overview of

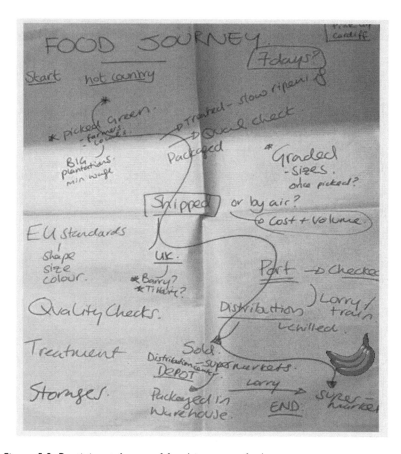

Figure 5.2 Participants' map of food journey of a banana.

the FSC, for instance country of origin and key steps in the supply chain, there was a general haziness regarding the details of each step and sometimes where foods came from; thus bananas come from '*Tenerife or somewhere*' or just '*overseas*' (Glasgow group 4 60+y, all social grades). There was a general consensus that FSCs are long and complex, as illustrated in the maps (see Figure 5.2).

The concept of a FSC itself was, however, for most people, abstract and opaque. Indeed, some in urban areas made explicit references to a sense of a loss of connection to FSCs:

> *I've become distant from the food chain simply because I live in central London . . .*
> *living in the middle of a city . . . that connection to the origins of food gets lost.*
>   London group 1 60+y
>   Any social grade

However, many simply noted that where their food came from was something they neither knew, nor particularly cared, about:

> I've never thought about it much.
>> Glasgow group 4 60+y
>> All social grades

> I'm really not that bothered where it comes from, my criteria is the end flavour, and how good it tastes.
>> Cardiff group 2 25–59y
>> Social grades ABC1

Despite limited knowledge of, or even concern about, the detail of FSCs, participants across all groups made a number of considerations and assumptions. Firstly it was assumed that FSC processes are primarily driven by cost and time, which in turn affect food quality and taste. Retailers and food suppliers were thought to seek out cheap products and quick supply chains, abiding by only the minimum quality standards, in favour of achieving higher financial profit. Battery farming of chicken was an example provided by participants as a cheap and quick approach compared with free range farming, which was presumed to have a FSC with higher costs. Participants held the view that food suppliers and retailers may exploit consumer concerns about particular foods by overpricing foods from certain FSCs, such as organic, seasonal or local produce. This was considered unfair as it may exclude consumers on lower budgets from making what were seen as 'healthy' purchases.

While participants had little awareness about food safety regulations and quality checks that might occur along the FSCs for different commodities or who might perform these, they nonetheless assumed that with the FSC, quality and safety checks did occur:

> We are a very trustworthy nation. . . . I don't think we have a choice. . . . When it comes to processed foods, you're at the mercy of the manufacturers . . . you just trust that it's arrived there through a supply chain that has protected it to a quality level . . . once we start to doubt it we'd all have nervous breakdowns!
>> London group 1 60+y
>> All social grades

This assumption was evidenced by the empirical observation that generally people are not unwell after consuming food they purchase. Some participants believed that food standards and regulations are of a higher quality in the UK and the European Union than elsewhere in the world. Participants wondered how checks were placed on food produced outside the UK, where these assumptions that 'someone' was looking out for quality and safety were less secure. Food with origins closer to home was therefore considered to be safer and better, particularly if sourced from known local retailers who were assumed to have shorter FSCs, or if grown oneself, so the FSC was known in detail:

> I would have more confidence in a butcher than buying [meat] in a supermarket . . . a butcher knows what a good meat is . . . it's also handled by less people.
>   Cardiff group 2 25–59y
>   Social grade ABC1

> If you grow it yourself, you know what's gone into it.
>   Bradford group 3 25–59y
>   Social grades C2DE

Everyday management of concerns about the FSC

Generally, then, where food came from was not a front of mind issue for most participants. Knowledge (or lack of it) around the FSC did not impact significantly on decisions around choosing and buying food. These were instead described as everyday routine practices, in which overt considerations about FSCs were generally limited.

> I don't really think about where it's from.

> You do the best you can.
>   Plymouth group 2 25–59y
>   Social grades ABC1

> If I like I eat it.
>   Glasgow group 4 60+y
>   All social grades

In terms of accounting for how they made decisions about what to buy and what to avoid, participants described the same 'rules of thumb' that have been widely reported in other studies of food safety, such as sensory evidence and habitual choices:

> I think you should use your nose. Have a good sniff at it.
>   Glasgow group 4 60+y
>   All social grades

> My approach is that if it looks OK and smells OK, then it probably is OK.
>   Cardiff group 2 25–59y
>   Social grades ABC1

Some reported that food label information, such as country of origin, ingredient lists and the Fair Trade logo, provided short-hand information about the FSC and an assurance of quality and safety. Others, however, did not use label information at all.

> I don't really look at it [labeling] to be honest.
>   Plymouth group 2 25–59y
>   Social grades ABC1

Given that decision making was largely routinised and habitual, at times participants struggled to articulate the criteria that affected food purchasing decisions, and how the FSC might factor in those; it was simply not something that many actively thought about. Considerations around the FSC were not clearly distinct or separable from other factors that affected choices around food and participants' narratives described the range of criteria they had to negotiate when making food purchasing choices, such as cost versus quality and health concerns, taste and time constraints. Many of these criteria involved explicit tradeoffs; a commonly reported one was that of cost and quality, which were seen as inevitably inversely related:

> We would all love to shop at Waitrose, but it doesn't work that way.

> If you can afford it, you'll get the proper stuff.
>   Glasgow group 4 60+y
>   All social grades

An exception to the routine background invisibility of FSCs was 'food scares', such as the horsemeat crisis. For some, the horsemeat crisis in 2013, in which products labelled as containing beef had been found with horsemeat,[4] had undermined their trust in food labels. All participants reported that media coverage of these scares would trigger concerns about food quality and safety, and potentially active seeking of information and consideration of where food comes from. Some participants could recite crises from several years earlier, and tended to remember them in greater detail if they felt affected by it personally. For example, women in the older age groups recited consumer alerts about baby milk formula in the 1970s, which resulted in long-term behaviour change for them, as they stopped using particular brands of formula to protect their children. More recent examples, such as the horsemeat, two years prior to the fieldwork still elicited strong reactions:

> Disgusted. . . . If you're buying a beef burger, you expect it to be beef and then when you find out on the news that it's horse you think 'Ugh, that's disgusting', you're not going to buy it again are you?
>   Plymouth group 1 18–24y
>   All social grades

This expectation that a beef burger should contain beef suggests something of the routine trust in FSCs in general. The assumption was that they were reliable, and that this reliability was ensured by a generalised, if unknown, governance. The media played a key role in breaching this generalised trust, prompting concerns about the FSC. This ranged from news stories about errors within the FSC (e.g. the horsemeat crisis and BSE) or outbreaks of food related ill-health (e.g. *E. coli* or *Salmonella*) or changes to FSC regulations, but also included TV documentaries explaining or providing insight about the FSC processes, investigative journalism and exposés of poor FSC practices. TV chefs also had a powerful voice for supplying information to participants about FSC via food shows. In addition, internet and social media were important sources. For example, participants might view Facebook posts regarding an aspect of the FSC, which may result in them conducting further online investigation, via a range of web-links.

I wouldn't go looking for it, but if it came up on the news I would read it.

I haven't eaten KFC since I saw that video about the way they keep their chickens and the conditions – it was disgusting . . . it kind of makes you go 'oh, I don't know if I want to eat that anymore' . . . you actually realise what you're eating.
   Belfast group 1 18–24y
   All social grades

In the absence of a food scare or other trigger many people were thus largely disinterested in where their food came from, and FSC considerations were not (routinely) part of their everyday decision making. This is not, perhaps, surprising in the light of theories of modernity, suggesting that routinised trust enables a lack of detailed understanding of complex modern institutions, until something breaches their taken-for-granted ordinariness. However, what was perhaps more surprising was that, for some participants, there was a more explicit orientation to ignorance: an active form of not-knowing.

I wouldn't delve into it too much.
   Plymouth group 2 25–59y
   Social grades ABC1

If you start thinking about it you wouldn't eat a thing.
   Glasgow group 4 60+y
   All social grades

I'd rather not know . . . you can't know everything. It would put you off some stuff.
   Belfast group 2 18–24y
   All social grades

Such ignorance was reported as an active strategy of management: a deliberate decision to 'not know' about the FSC. 'Not-knowing' enabled participants to procure food, prepare it and eat it without undue concern.

Trust and responsibility

Given that consideration of FSCs was not an everyday concern for many participants, attributions of trust and responsibility were largely framed in relation to abstract other agents, such as 'government' or 'supermarkets'. These generalised parties were relied on to ensure that, within the food system, there would be adherence to minimum levels of standards and regulations. Different levels and types of responsibility and trust were seen as residing in different actors, who were attributed as having particular roles and motivations for being involved in the chain.

Government bodies

Government bodies were ultimately seen as responsible in ensuring that quality standards are upheld, as they were seen as acting in the public interest rather than financial gain. While seen as abstract and distant, they were more trusted than most private sector actors, such as retailers and producers.

> . . . should be some sort of government body.
> Glasgow group 4 60+y
> All social grades

There was an appreciation that government bodies cannot check every food item available to consumers, but they were seen to perform a vital role in setting minimum standards and working together with producers and retailers to ensure these are met. There was also an expectation that government bodies perform spot checks on food producers and retailers, and had the power to sanction poor performing organisations. The threat of such sanctions was considered enough to ensure most retailers and supermarkets would maintain the expected standards as a matter of course.

Food producers

Food producers were mostly considered to be profit driven, and as being pressurised by big supermarkets to further lower the cost of their products. Participants therefore envisaged short-cuts might be taken to reduce production time and/or costs. Time constraints were seen as linked to a decrease in food quality and increased risk of harm to food products and consumers. Food producers were therefore trusted to comply with only the minimum required government standards, as going beyond these might add to their time and costs.

End retailers

End retailers, such as supermarkets, shops and restaurants, were considered to be responsible for food safety and that foods are accurately labelled and safe to eat at point of purchase. There was an expectation that they should not knowingly sell food to people that is mislabeled or unsafe. End retailers were considered to be motivated mainly by profit, and so they were trusted less than government bodies. However, participants believed that this also worked in favour of consumers in that retailers would maintain a certain level of food standards and safety along the FSC to uphold a good reputation among consumers. Retailers were also expected to place intermittent checks on their food suppliers and distributers.

Participants trusted that big retailers and supermarkets would perform adequate food safety and quality checks as a form of insurance, to avoid consumers becoming ill and possible resulting 'bad news stories' or costly lawsuits. This attribution of reputation as a driver for quality assurance was evident in two particular examples: high-end supermarkets and local retailers – or, more specifically, typically the figure of the 'local butcher':

> *A company like M&S will be very careful.*
> *Plymouth group 2 25–59y*
> *Social grades ABC1*

> *In the supermarkets, they don't care what the quality's like but the local butchers and the local green grocer would try and source the best [food] to make sure you go back again.*
> *Belfast group 4 60+y*
> *All social grades*

Thus there was a level of routine scepticism about the motivations, if not the trustworthiness, of big corporate food retailers and supermarkets and participants had greater trust in small and specialist retailers, such as butchers, fishmongers and greengrocers. There was a notion that small businesses have a closer connection to the source of the items they sell and take greater care in sourcing each product and possibly even have personal, friendly relationships with the food producers and distributers. This impression was rooted in the assumption that small retailers have to work harder to ensure they retain their customers.

Trust was therefore bestowed on high-level actors assumed to be working for the public and not driven by profit (such as government agencies and food inspectors) and in local, known retailers, assumed to be driven by consumer reputation, which was assumed to be easily damaged if deficiencies in the FSC were allowed.

Consumers

> *The ideal is that we buy seasonal foods but the public expect food to be available all year round.*
> *London group 1 60+y*
> *All social grades*

Participants saw their own role as less clear-cut. Consumers were seen to be responsible for the food choices they make and the impact of these on the FSC. Consumers were seen to create demand for out-of-season foods all year round, but the supermarkets were also considered to be complicit in this change to the FSC, as they provide out-of-season foods, which may have greater carbon footprints and air-miles, for example.

Participants also considered that consumer choices are generally dictated by cost, convenience and marketing. They therefore felt that other bodies within the FSC had a responsibility to protect consumers and help them make informed choices. For example, supermarkets and food companies were expected to have information available about the FSC for interested customers, on their websites for example.

Discussion

Food publics

Overall these data show that for many people FSCs, and the larger food system, were mostly rather abstract concepts. This echoes the findings of others that terms such as 'food security' are opaque, and the food system is seen as a black box (Darnton, 2016; Dowler et al., 2011). However, people's views and ways of engaging with the FSCs were not homogenous, and we could identify three broad styles of engagement. First, there were those who were highly engaged and aware. These participants pro-actively sought out information and made conscious choices to eat/not eat certain foods because of their particular political, ethical and/or religious values or health condition. Second were those who were routinely disengaged and less well informed, but who when prompted would seek information and change their practices, for instance in response to a food scare and/or media report. Third were participants who reported that they 'don't know and don't care'. These participants had not only limited information about FSCs, they also expressed little interest in it. Rather, their primary concern was the end food product or meal. This lack of concern was most typical for those with little involvement in shopping or food preparation. Amongst this group were those who actively did not want to think about the supply chains for some foods, such as battery chickens. Thus, as Darnton (2016) argues, while there is tendency to think of the public or consumers as a single entity, and particularly among policymakers, rather there are diverse food publics and these cannot be predicted by the socio-demographic indicators of age, gender or social status.

Not-knowing

The views reported here also resonate with other studies that have examined public perceptions of food-related risks in that risk and anxiety were not the primary tropes in which people described their concerns about the FSC, and that tacit routinised rules of thumb played a key role in people's everyday decision making about food purchases (Green, Draper, and Dowler, 2003; Green, 2009). However, what has been under-reported in other studies is the role of 'not-knowing'. We found this to be a rather more complex, and multiple, state than has previously been identified. There were different kinds of not-knowing with different implications for how people engage with FSCs, and different implications for policy. Agnotology, or the study of ignorance, is now a growing field (Smithson, 1989), although much recent research focuses on the study of the role of ignorance at the systemic or organisational level (Mair, Kelly, and High, 2012) and how vested interests may be deliberately fostering doubt and uncertainty as a political goal (Proctor and Schiebinger, 2008; McGoey, 2012). The functions of ignorance in everyday life, and the ways in which it might be a rational strategy of management, have been perhaps under-explored. As Croissant (2014) argues, we need to differentiate between forms of 'ignorance' or non-knowledges; also that these are not necessarily an absence of knowledge or information.

In this study, three different forms of not-knowing or non-knowledges emerged. Firstly, not-knowing as a form of strategic ignorance that allows the accomplishment of everyday life and routine decision-making in the face of complexity and competing concerns. Crucially, however, this kind of not-knowing relies upon trust. This finding is similar to de Krom and Mol's (2010) study of avian influenza in the UK, which found that trust allowed people to bracket not-knowing and enabled everyday food purchasing practices whether from both known and small-scale food system actors or large-scale and impersonal retailers. This form of not-knowing is enabled by trust and should not be underestimated or assumed to be a negative quality in need of correction. Rather as Simmel wrote, it is a condition that enables the practice of everyday life: 'the hypothesis of future conduct, which is sure enough to become the basis of practical action, is, as hypothesis, a mediate condition between knowing and not-knowing' (Simmel, 1909, p. 450).

In other domains, a level of ignorance or non-knowing about the detailed operations of a system has also been identified as crucial to enabling trust, which is easily damaged by, for instance, over-production of information for the public on quality indicators or other system outcomes (see e.g. Brown, 2008; Legido-Quigley et al., 2014). 'Ignorance' or not caring to know can thus be an important indicator of the trustworthiness of the public sphere: that people do not care suggests they trust the systems of governance that are in place. Forcing knowledge upon those publics risks undermining faith in the system as whole, rather than improving it. Of course this is not to argue that truths should be deliberately hidden from the public, but, as the philosopher Onora O'Neill argued in her 2002 Reith Lectures, perversely an excess of transparency might not only not build trust but actually undermine trust in that it removes the conditions for not-knowing. She argues that deception and misinformation are the real enemies of trust.

The second form of not-knowing was to avoid knowledge of uncomfortable truths, and particularly the suffering of animals in the production of food. This is ignorance as a form of denial that allows uncomfortable truths to be made invisible and unrecognised (Wicks, 2011). It resonates with Jackson et al.'s study of the moral economy of different foods (sugar and chicken) in which they found that that there were areas of silence and 'collective amnesia' for both foods and that 'the process of selective remembering and forgetting articulates moral issues [across time]' (Jackson, Ward and Russell, 2009, p. 16).

The third form of not-knowing was the eschewing of rational knowledge as relevant to the business of food related choices. Here, instead, other forms of knowledge based on sensory data, experiential learning and, following Rayner (2012), specifically the unconscious heuristics or rules of thumb that people used to make decisions and to judge the safety of food were prioritised. These are the forms of 'non-rational' knowledge that have been rendered less legitimate in modernity, yet are still crucial to decision making: these become positioned as 'ignorance' and in need of correction, an assumption based on the discredited 'deficit model'.

Ignorance and making the public

The ways in which the public have been constructed as the objects of policy has shifted historically from the late 19th century to present (Draper and Green, 2002). Threading across many of these constructions is the assumption that the public are ignorant and that this explains negative attitudes towards new technology, such as GMOs and synthetic biology. It therefore follows that redressing this ignorance will 'correct' such attitudes and 'bad' behaviours (Marris, 2015). Latterly the public have more often been constructed as consumers, and now citizens with responsibilities, and as obligated to know and act in certain ways (Evans, Welch, and Swaffield, 2017). However, our findings suggest that not everyone wants to be a food citizen. More significantly, these findings also suggest that 'ignorance' is not necessarily a deficit.

Implications for methods

As Last writes 'not-knowing and not-caring-to know are genuine attitudes of mind' (Last, 1981, p. 387) and something that we should perhaps pay more attention to. The role of qualitative research is often seen as creating knowledge about the knowledge of others, but studying absences presents considerable methodological challenges (Croissant, 2014). A common finding on research on many topics is that participants neither know much about the topic, nor really care very much about the topic. As Last (1981, p. 387) noted, 'every investigator has received the answer "don't know" and been unsure whether the answer was the truth or simply a snub' – as he says, it is hard enough to find out what people know, let alone what they don't know. Researching 'non-knowledge' is particularly problematic because our methods of qualitative inquiry are designed to generate knowledge, and participatory methods such as these deliberative workshops more than most. In this study deliberative methods were chosen to access people's tacit knowledges, but it was the visualisation exercises in which participants collectively mapped food supply chains that literally revealed the blanks.

Notes

1 Corresponding author.
2 The study received ethical approval from NatCen's Research Ethics Committee in 2014 (Ref: I11052). NatCen has an ethics governance procedure that meets the requirements of the UK Economic & Social Research Council (2005) Research Ethics Framework and the UK Government Social Research Unit's (2005) GSR Professional Guidance: Ethical Assurance for Social Research in Government, Cabinet Office, London.
3 In this classification A, B, C1 are used to describe higher, non-manual and skilled occupations; C2, D, E to describe those in manual and low-grade occupations or unemployed.
4 The horsemeat crisis of 2013 was caused by the revelation that some beef products, such as beef burgers sold in several British and Irish supermarkets, contained undeclared horse meat. The scandal later spread to other European countries, and it exposed both the complexity of and lack of traceability in the beef supply chain.

References

Beardsworth, A. and Keil, T. (1997). *Sociology on the menu: An invitation to the study of food and society*. London and New York: Routledge.

Beck, U. (1992). *Risk society: Towards a new modernity*. London: Sage.

Brown, P.R. (2008). Trusting in the new NHS: Instrumental versus communicative action. *Sociology of Health and Illness*, 30(3), pp. 349–363.

Caplan, P. (2000). Eating British beef with confidence: A consideration of consumers' responses to BSE in Britain. In: P. Caplan, ed., *Risk revisited*. London: Pluto Press, pp. 184–203.

Croissant, J.L. (2014). Agnotology: Ignorance and absence or towards a sociology of things that aren't there. *Social Epistemology*, 28(1), pp. 4–25.

Darnton, A. (2016). *Our food future: Literature review*. London: UK Food Standards Agency.

de Koning, K. and Martin, M. (1996). Chapter 1: Participatory research in context: Setting the context. In: K. de Koning and M. Martin, eds., *Participatory research in health: Issues and experiences*. London and Atlantic Highlands, NJ: Zed Books Ltd, pp. 1–18.

de Krom, M.P.M.M. and Mol, A.P.J. (2010). Food risks and consumer trust: Avian influenza and the knowing and not knowing on UK shopping floors. *Appetite*, 55, pp. 671–678.

Dowler, E., Kneafsey, M., Lambie, H., et al. (2011). Thinking about "food security": Engaging with UK consumers. *Critical Public Health*, 21(4), pp. 403–416.

Draper, A. and Green, J. (2002). Food safety and consumers: Constructions of choice and risk. *Journal of Social Policy and Administration*, 36(6), pp. 610–625.

Evans, D., Welch, D. and Swaffield, J. (2017). Constructing and mobilizing "the consumer": Responsibility, consumption and the politics of sustainability. *Environment and Planning*, 49, pp. 1396–1412.

Food and Agriculture Organization (2014). *Sustainable diets and biodiversity*. Rome: Food and Agriculture Organization.

Food Standards Agency (2016). *Our future food*. London: UK Food Standards Agency.

Foresight (2011). *The future of food and farming: Challenges and choices for global sustainability*. London: Government Office for Science.

Giddens, A. (1991). *Modernity and self identity*. Oxford: Polity Press.

Gofton, L. (1990). The rules of the table: Sociological factors influencing food choice. In: C. Ritson, L. Gofton and J. McKenzie, eds., *The food consumer*. Chichester: John Wiley and Sons Ltd.

Green, J. (2009). Is it time for the sociology of health to abandon "risk"? *Health, Risk and Society*, 11(6), pp. 493–508.

Green, J., Draper, A.K. and Dowler, E. (2003). Short cuts to safety: Risk and 'rules of thumb' in accounts of food choice. *Health, Risk and Society*, 5(1), pp. 33–52.

IPES-Food (2017). *Unravelling the food and health nexus: Addressing practices, political economy, and power relations to build a healthier food system*. The Global Alliance for the Future of Food and IPES-Food.

Jackson, P. (2015). *Anxious appetites: Food and consumer culture*. London: Bloomsbury.

Jackson, P., Ward, N. and Russell, P. (2009). Moral economies of food and geographies of responsibility. *Transactions of the Institute of British Geographers*, New Series, 34, pp. 12–24.

Kneafsey, M., Venn, L., Holloway, L., et al. (2008). *Reconnecting consumers, producers and food: Exploring alternatives*. Oxford: Berg.

Last, M. (1981). The importance of knowing about not knowing. *Social Science and Medicine*, 15B, pp. 387–392.

Legido-Quigley, H., McKee, M. and Green, J. (2014). Trust in health care encounters and systems: A case study of British pensioners living in Spain. *Sociology of Health & Illness*, 36(8), pp. 1243–1258.

Mair, J., Kelly, A.H. and High, C. (2012). Introduction: Making ignorance an ethnographic object. In: C. Hig, A.H. Kelly and J. Mair, eds., *The anthropology of ignorance*. New York: Palgrave Macmillan, pp. 1–32.

Marris, C. (2015). The construction of imaginaries of the public as a threat to synthetic biology. *Science as Culture*, 24, pp. 83–98.

McGoey, L. (2012). Strategic unknowns: Towards a sociology of ignorance. *Economy and Society*, 41(1), pp. 1–16.

New Citizenship Project (2017). *Food citizenship: How thinking of ourselves differently can change the future of our food system*. Available at: https://drive.google.com/file/d/0B0swicN11uhbSGM2OWdCeXdQZGc/view

O'Neill, O. (2002). *A question of trust*. The 2002 Reith Lectures. London: BBC.

Proctor, R.N. and Schiebinger, L. (2008). *Agnotology: The making and unmaking of ignorance*. Stanford, CA: Stanford University Press.

Rayner, S. (2012). Uncomfortable knowledge: The social construction of ignorance in science and environmental policy discourses. *Economy and Society*, 41(1), pp. 107–125.

Ritchie, J., Lewis, J., Nicholls, C.M., et al. (eds.) (2013). *Qualitative research practice: A guide for social science students and researchers*. London: Sage.

Simmel, G. (1909). The sociology of secrecy and secret societies. *The American Journal of Sociology*, XI(4), pp. 441–498.

Smithson, M. (1989). *Ignorance and uncertainty: Emerging paradigms*. New York: Springer Verlag.

Unger, S. (2003). Moral panics versus the risk society: The implications of changing sites of social anxiety. *British Journal of Sociology*, 52(2), pp. 271–291.

Wicks, D. (2011). Silence and denial in everyday life – the case of animal suffering. *Animals*, 1, pp. 1866–199.

Chapter 6

Eating less meat 'to save the planet'

Studying the development of sustainable healthy eating advice in the UK and Denmark

Isabel Fletcher

Introduction

In the last ten years, the topic of sustainable diets – how we should eat in order to minimise environmental harms – has gone from being a specialist concern to the subject of mainstream news reporting and high-level policy documents. This new prominence can be seen as part of wider processes of 'convergence' (Friedmann, 2017) whereby international (and perhaps national) policy increasingly recognises the interrelationships among health, agriculture and environment. Discussions of the multiple different kinds of harm resulting from the way we eat and how to reduce them necessarily involves making new linkages between policy areas – for example, food production, public health and environmental pollution – that have historically been managed in distinct and well-separated policy areas or siloes.

The environmental impacts of food production and consumption – often now labelled the food system (Lang, Barling, and Caraher, 2009) – have been recognised since the beginnings of the modern environmental movement (Carson, 1965; Lappé, 1973). In its work on sustainable development, the United Nations World Commission on Environment and Development – better known as the Brundtland Commission – highlighted the ways in which intensive food production has led to deforestation, desertification and pollution (UNWCED, 1987). This body introduced the concept of sustainable development into global policy discourses. Since the 1992 Earth Summit in Rio there have been a series of UN conferences on sustainable development. In parallel, sustainability science has developed into an important interdisciplinary research area that retains roots in environmental sciences (Kates, 2011).

Over the same period, a parallel set of arguments have been developed about the harms caused by food consumption, based on biomedical research and especially nutrition. The negative health impacts of modern diets on health have been recognised since before the 1980s (Walker and Cannon, 1985; Lang and Heasman, 2004) and have now become a part of mainstream public health nutrition and food policy (The Lancet, 2016; WHO Europe Office, 2014). This is largely because global increases in rates of chronic diseases – such as heart disease, diabetes and stroke – are seen to be caused by diets containing too many processed foods high in fat, salt and sugar (WHO, 1990; Popkin, 2004; Monteiro et al., 2017).

However, discussing sustainable diets involves making new connections between two sets of evidence: the health concerns of public health nutrition and food policy campaigners, and evidence about the impact of the food system on greenhouse gas emissions, water and air pollution, biodiversity and soil degradation. The term 'sustainable diets' has only been widely used since 2010. However, it dates back to the mid-1980s when Joan Dye Gussow and Katherine L. Clancy – two American academics based in departments of nutrition education and human nutrition respectively – published 'Dietary Guidelines for Sustainability' (Gussow and Clancy, 1986; see also Gussow, 1999). They argued that 'consumers need to make food choices that contribute to the protection of our natural resources' (quoted in Gussow, 1999, p. 194). Their approach does not seem to have been widely adopted until 2010 when the UN Food and Agriculture Organization (FAO) and Biodiversity International organised a symposium on biodiversity and sustainable diets. Gussow and Clancy's article has now been cited nearly 200 times, and is acknowledged as an important forebearer in the literature on sustainable diets (Mason and Lang, 2017).

Definitions of sustainable food systems and diets[1] are necessarily complex. They need to incorporate multiple and overlapping criteria that assess the environmental, health, social and economic aspects of the food we produce and consume (Mason and Lang, 2017). This complexity, combined with the newness of the topic, means that no specific definitions of sustainable diets have been produced, either for particular populations or for regions. In its report on the 2010 symposium, the FAO produced the most widely cited definition. This is better described as a set of broad principles, rather than a guide to how to eat:

> Sustainable diets are those diets with low environmental impacts which contribute to future food and nutrition security and to healthy life for present and future generations. Sustainable diets are protective and respectful of biodiversity and ecosystems, culturally acceptable, accessible, economically fair and affordable; nutritionally adequate, safe and healthy; while optimizing natural and human resources.
>
> (FAO, 2012, p. 7)

Like the 1947 World Health Organization (WHO) definition of health, this definition is hugely aspirational and widely cited, but also gives no guidance on the practicalities of achieving a sustainable diet in particular contexts. It seems to have marked the beginning of a new policy interest in the topic, but it is not yet embedded in international policy and practice to anything like the same extent as the FAO's definitions of food security (Shaw, 2008). This may be a matter of time – the concept of food security was first used by the FAO in the early 1970s and has been revised several times since (Maxwell, 1996). Alternatively, it may point to the ways in which the still-evolving discourse of sustainable diets is less securely 'owned' by the FAO, due to its origins in research and activism rather than development programmes.

However, the UK branch of the World Wide Fund for Nature (WWF) has developed the Livewell Principles (see Figure 6.1) from its project of the same name. The WWF is a large conservation-focused environmental charity, and WWF-UK's innovative One Planet Food Programme (2009–12) set goals to reduce UK food-consumption-related emissions by at least 25 percent by 2020 (Macdiarmid et al., 2011). This was followed by the Livewell for LIFE project (2011–14) which, in partnership with the Friends of Europe think-tank, focused on sustainable diets in France, Spain and Sweden (Livewell for LIFE, 2014). Due to variation in food production systems as well as incomes, eating cultures and average diets, it is difficult to specify precisely what a sustainable diet would look like in different contexts – even if that context is limited to three high-income European countries. Therefore, the Livewell Projects developed a set of broad principles to guide individuals, rather than specific rules.

In what follows, I will first outline the theoretical framework for my ongoing research on sustainable diets, which combines approaches from Science and Technology Studies (STS), regulation theory and policy studies. I will continue by describing the literature review I undertook at the beginning of this project,

Figure 6.1 The Livewell Principles.
Source: WWF (2017, p. 10).

and discuss some of its key findings. I then describe the rationale for interviewing expert stakeholders in Denmark and the UK about their understandings of sustainable diets, and how/whether sustainability criteria could be incorporated into existing dietary guidelines.

Theoretical/analytic framework

Food policy is regularly described as complex; issues such as food security and obesity are seen as 'wicked problems' requiring solutions developed out of wide-ranging multi- or inter-disciplinary collaborations (Foresight, 2007a, 2011). Partly this complexity arises from the ways in which food issues span many domains, including (but not limited to) environment, health, social welfare and trade. However, historically, different aspects of food policy have been conducted by specific ministries within national governments. In the period since the Second World War, the UK Department of the Environment Food and Rural Affairs (DEFRA) and its various predecessors, has been responsible for agriculture and food production policies, whereas the Department of Health has been responsible for nutrition policies. This institutional arrangement has led to the development of well-entrenched policy siloes (Lang, Barling, and Caraher, 2009), meaning that policies often focus on either production (agriculture, investment and trade) or consumption (health and social welfare), and only in newer areas such as obesity and sustainability have there been attempts to address both aspects of the food system (Foresight, 2007b; Garnett, 2014). One aim of my research is to investigate whether by making demands for the incorporation of environmental criteria into healthy eating guidelines, the development of government-mandated models of sustainable eating will contribute to the breaking down of such siloes. Such policy integration is seen as a necessary step in solving both diet-related ill health and the negative environmental effects of the food system (Barling, Lang, and Caraher, 2002; Lang and Barling, 2012).

Recent work within regulatory theory also provides approaches that help to understand processes within contemporary global food systems. Black (2001) has developed a model of 'de-centred regulation', a means of going beyond traditional 'command and control' models that understand regulation as a matter of legal rules produced by governments and enforced by means of criminal sanctions:

> Decentring is . . . used to express the observation that governments do not, and the proposition that they should not, have a monopoly on regulation and that regulation is occurring within and between other social actors, for example large organizations, collective associations, technical committees, professions etc., all without the government's involvement or indeed formal approval: there is 'regulation in many rooms'. Decentring is also used to describe changes occurring within government and administration: the internal fragmentation of the tasks of policy formulation and implementation.
>
> (Black, 2001, pp. 103–104)

The outcomes of such decentring processes include information asymmetries between regulators and regulates resulting from the complexity of interactions within the regulatory system, and the collapse of the public–private distinction in governance and regulation (Black, 2001).

As the food industry has expanded in the last 40 years, its complex supply chains are increasingly controlled by large private corporations with minimal oversight by governments (Lang, Barling, and Caraher, 2009). A lack of government control over these systems has led to major information asymmetries, as highlighted in the 2013 European horsemeat scandal (Lawrence, 2013), which showed that neither national governments nor large retailers had complete information about the complex supply chains for budget frozen burgers. Demonstrating their autonomy, food companies have also developed their own regulatory systems, such as Hazard Analysis and Critical Control Point (HACCP) procedures, which have, in turn, been taken up by the WHO and FAO in their joint work on food safety (Lang, Barling, and Caraher, 2009). Finally, blurring the public–private distinction, there are an increasing number of regulatory schemes that have been created by the food industry in partnership with governments, NGOs and other bodies. In the UK, examples include the Assured Food Standards (AFS) scheme developed by the National Farmers Union, and the Marine Stewardship Council that is a partnership between Unilever and the World Wildlife Fund (Lang, Barling, and Caraher, 2009).

Both food systems approaches and this branch of regulatory theory describe the complexity of multi-level food governance systems and highlight the variety of actors – many outside of government – who are contributing to technical knowledge about sustainable diets and suggesting potential regulatory solutions. Analysing the ways in which scientific research interacts with policymaking is a key concern of science and technology studies (STS). Researchers working in this field, including Sheila Jasanoff (2005), Brian Wynne (1996) and Steve Yearley (2000), have produced key empirical case studies replacing linear and technocratic accounts of policy based on 'sound science' with more sophisticated narratives. These describe complex processes of problem framing and the allocation of expert authority in, for example, the regulation of air pollution, pesticides and radiation levels. In such cases, researchers need to undertake boundary work (Gieryn, 1999), navigating shifting boundaries between the 'technical' and the 'political' in order to provide credible advice. Incorporating sustainability criteria into existing healthy eating advice will raise questions about who counts as an expert on these matters, and what kinds of evidence should be used to assess the environmental impacts of different foods and patterns of eating. Such questions are likely to be contentious.

Since the early 20th century, nutrition research evidence has been used to create dietary advice. Nutrition science has thus become an influential part of modern daily life. Despite its pervasiveness and (presumed) influence, compared to other areas of biomedicine, nutrition science remains relatively understudied by social science researchers. There are studies of the development of specific versions of the US dietary guidelines. These include Garrety's (1997) account of the development of low-fat healthy eating guidelines and Hilgartner's (2000) account of the controversy surrounding the production of the US Government's 1995 Dietary Advice for Americans. Both these accounts demonstrate the ways in which technical elements of nutrition research can become highly politically contentious when used to develop official guidelines. British historians of medicine have also analysed the work of several important advisory committees that produced official dietary advice (Bufton, 2005; Bufton and Berridge, 2000; Smith and Bufton, 2004). These accounts illustrate shifting boundaries between technical and political issues as researchers sought to resolve repeated controversies about the necessary amounts of key dietary components – such as protein, sugar and animal fat – as causes of ill health. These controversies are still part of debates about dietary advice (LaBerge, 2008; Johns and Oppenheimer, 2018).

As part of debates around sustainable diets, it has been suggested that official dietary guidelines should be extended beyond their focus on healthy eating to include advice about the sustainability of specific food items (often animal products) and also patterns of eating (Mason and Lang, 2017). This provides a case study in the formation of new food regulation and a new interdisciplinary body of research evidence combining nutrition and sustainability science. It also enables researchers to study the ways in which this new body of knowledge is then translated into public-facing and relatively straightforward guidelines for individuals and institutions. Accounts from nutrition researchers (e.g. Jahns et al., 2018) describe a process based on medical models of systematic review where research evidence is carefully gathered, assessed and ranked according to standard criteria. Such processes have been designed to handle specific kinds of evidence – quantitative data framed in terms of causal relationships and biological pathways. It is difficult to envisage how evidence pertaining to the wider aspects of sustainability, in particular its social and political elements, can be incorporated into dietary guidelines.

Sustainability, as I have argued, is a complex concept, but nutrition research is also highly complex. The physiological processes underlying eating are enormously varied and complicated. Diets are long-term patterns of eating – study over weeks, months or even a couple of years does not fully capture their effects on health. They are also very varied – people eat a large number of different foods and very many combinations of those foods are compatible with health (Nestle, 2002). Social science research has shown that food choice is governed by many factors other than health or environment. In the UK price and familiarity/branding dominate food purchasing decisions (DEFRA, 2016) but overarching factors such as age, gender, ethnicity and class also have a profound influence on food preferences (Delormier, Frolich, and Potvin, 2009). Finally, food policy is politically contentious – the food industry is large and powerful and prepared to influence policymaking when its interests are threatened, by, for example, official recommendations to reduce red meat consumption (Garrety, 1997; Freidberg, 2015). All these factors shape the creation of dietary guidelines and make the integration of sustainability criteria a complex and uncertain process.

Literature review

In the initial stages of this research, I undertook a review of the existing academic and policy literature on sustainable diets (Table 6.1). To search the academic literature, I conducted trial searches on several databases including ASSIA (Applied Social Science Index and Abstracts), IBSS (International Bibliography of Social Sciences), Medline, Scopus and Web of Science. Ultimately, I decided to use Scopus because it was the only search engine that provided specific and usable results across several research areas – Medline does not include environmental science research journals, and a Web of Science search using the same terms and limits produced more than 2500 results, few of which were directly relevant.

I conducted a combined search for the phrases 'sustainable diet' and 'sustainable diets' using quotation marks and a wild card character. In order to capture peer-reviewed publications I limited my search to articles and reviews. On the basis of preliminary searches, I also limited the time frame to the last 15 years.

The top five subject areas were agricultural and biological sciences, nursing, medicine, environmental sciences and social science and limiting the search to these areas gave me 172 articles. I then excluded pieces that dealt with the diets of non-humans, those that only mentioned the term 'sustainable diets' and, finally, those that did not discuss what was meant by the term beyond citing or quoting the FAO definition. At the end of this process, I was left with 85 articles, published between 2004 and 2018. As I only used one database, this is not a representative sample of the academic literature; rather, it is a snapshot that

Table 6.1 Literature review search strategy

Database: SCOPUS	**Search criteria:** 'Sustainable diet*' in abstract/title/ keywords	172
	Time period: 2003–present	
	Document type: Articles and reviews	
	Limits	
	Language: English	
	Subject areas: agricultural and biological sciences, nursing, environmental sciences, medicine (i.e. nutrition) and social science	
	Document type: article or review	
	Exclusion criteria	85
	Research on diet of non-humans (mostly aquaculture)	
	Abstract mentions sustainable diets in passing	
	Article contains no discussion of sustainable diet/s, beyond quoting the FAO definition	

can be used to highlight key features of this evolving discourse. In the rest of this section I will outline the results of this search, highlighting the journals in which articles appeared most frequently, their dates of publication, the location of the research outlined in these articles – both the institutional affiliations of the authors and where the data was collected – and finally, the methods used in the research.

The journals in which these articles were published give some sense of the main academic disciplines contributing to this discourse. The journals which appeared most often were *Public Health Nutrition* (13 articles), *Advances in Nutrition* (6), *Ecological Economics* (5), *Nutrients* (5) and *Food Policy* (4). Six out of the top ten journals were from nutrition and/or public health. *Public Health Nutrition* dominated the results with nearly 15 percent of the total articles, twice as many as the next title. Rather than sustainable diets being a topic that was shared equally between sustainability and nutrition research, it seems from these results that nutrition researchers are the main contributors to debates about the nature of sustainable diets and how to implement them. Many of these articles have several co-authors, and it is possible that such research is being conducted by multidisciplinary research teams that contain sustainability science researchers as well as nutrition researchers, but it is noteworthy that the research results appear to be largely reported in nutrition journals.

These results also confirm that this is a topic that has only recently begun to be of interest to researchers, and that this interest is growing rapidly. The first article in this search was published in 2004 (although, as discussed, the term 'sustainable diets' dates back to at least 1986). Between 2004 and 2010 there was about 1 article published each year. Since 2011, the number of publications has begun to increase, and reaches an annual total of 20 in 2016 and 22 in 2017. As 8 articles have already been published in first quarter of 2018, it looks like this growth in numbers will continue.

The majority of this research was produced by universities in western Europe and specifically specialised research institutes in the Netherlands, the UK and France. Exceptions were publications emanating from the FAO (largely concerning the sustainability of the Mediterranean diet), Deakin University in Victoria Australia and the NGO Biodiversity International. In terms of its geographical focus, research fell into two distinct categories – one which discussed sustainable diets at the global level without specifying a region, and another which specified the origin of the research data or the region under discussion. For this second category searching in English may have systematically skewed the results, despite its role as one of the major international languages of scientific research. However, bearing in mind this caveat, these results show that research into sustainable diets is largely conducted in the high-income countries of the world, especially in western Europe but also in Australia and the US. As well, those affiliated to academic institutions, other researchers publishing on this topic were employed by Blonk Consultants in the Netherlands that advises on life cycle analysis studies (discussed later in this section), the FAO, Biodiversity International (which is part of the UN) and the Soil Association.

More than half of the articles (44 in total) used data from or referring to these regions. Within these regions some countries were disproportionately represented – 11 studies were conducted in the UK and 8 in the Netherlands. This is due to the work of groups of academics in specialist groups such as the Institute for Environmental Studies (VU University Amsterdam) and the Rowett Institute (University of Aberdeen). With 4 articles, Sweden is also over-represented in the literature, which is probably due to its early adoption – even relative to the other Nordic countries – of sustainability criteria in its dietary guidelines (Roos, 2015). Other regions of the world were much less studied: 2 studies were conducted in Brazil, 1 in China, 1 in Qatar and 1 in South Africa, the only study that specified that it referred to a rural population.

These articles use an eclectic range of data sources, methods and theoretical approaches from nutrition and sustainability science, combined with those from other disciplines such as economics, policy studies, psychology, public health and sociology. There are three main types of articles in this collection: first, those that discuss (and sometimes compare) the environmental impacts of the diets of particular populations; second, those that discuss the environmental impacts of model diets, either ones they have created for the purpose or pre-existing model diets, such as the Mediterranean diet;[2] finally, those articles that review the literature in order to assess the existing evidence.

Important data sources for the first type of articles include national nutrition surveys – such as those conducted in the UK, France and Ireland – and life cycle analysis (LCA), which is often used to calculate the greenhouse gas (GHG) emissions associated with particular food items. LCA is the name given to a variety of techniques that model the life cycle, classically from 'cradle-to-grave', of products and services in order to assess their environmental impacts. Deriving from corporate studies of the environmental impacts of product packaging, LCA has achieved wider credibility in the last decade and its rules and procedures have been codified into International Standards Organization standards (Freidberg, 2016). Several of these authors (e.g. Jones et al., 2016; Auestad et al., 2015) acknowledge that LCA has become one of the most important ways of assessing the environmental impacts of the food system. In my sample, LCA data was mentioned or used in 21 of the articles, and was a key component of the evidence considered in review articles.

GHG emissions were the mostly widely used environmental indicator in these articles. Half of this sample (43 articles) referred to the greenhouse gas emissions of diets or specific foodstuffs, whereas only 8 discussed water usage, 7 land use and 5 biodiversity, and there was no discussion of wider sustainability issues such as animal welfare or workers' rights. GHG emissions are an imperfect indicator of the environmental impact of agriculture, as they do not capture other important effects, such as pollutions and resource degradation.

These limitations appear to be both well known and widely accepted (e.g. Ridoutt et al., 2017), yet they still appear to be the most widely used indicator in the literature.

In order to analyse the sustainability of diets, authors need to be able to combine complex sets of data from nutrition research and environmental research. A common approach, used both in studies that aim to understand the environmental impact of current diets and those that create models of less damaging ways of eating, is to combine nutrition research (to provide optimal diets or quantities of specific foods) and LCA (to assess the environmental impact of specific products) using linear programming. Statistical techniques, such as cluster analysis, can also be used to identify patterns within nutrition survey data, making it possible to compare the environmental impacts of these different eating patterns (e.g. Vetőné Mózner, 2014) or compare the impact of real-life and modelled dietary patterns (e.g. van de Kamp et al., 2018; Tyszler, Kramer, and Blonk, 2016). By means of these techniques, data on the environmental impacts of particular foods from sustainability science is becoming (partially) assimilated into nutrition science.

This rapid overview of the academic literature shows that research on sustainable diets is largely undertaken in western Europe by groups of researchers who often use life cycle analysis to measure the impact of food systems on GHG emissions and publish their research in nutrition science journals. However, as in other areas of nutrition and environment research, the academic and policy literatures on sustainable diets are closely related and frequently difficult to disentangle – key researchers write many different kinds of material including (but not limited to) academic articles, commentaries, literature reviews, meeting reports, policy reviews and reports on particular topics. In parallel to my search of the academic literature, I have also been collecting key reports and other publically available documents, some of which are listed in Table 6.2.

As it only includes English language publications, this list is obviously incomplete, even in a European context. However, it highlights the number of different organisations contributing to this discourse and is evidence of the important roles played by the FAO and other UN bodies in the development of scientific knowledge about sustainable diets. Other key contributors include large environmental NGOs such as WWF and Biodiversity International; departments and agencies of the Dutch, Swedish and UK governments; and academic networks, especially the UK-based Food Climate Research Network. This network is itself an example of the close relationship between academic and policy discourses on food and environmental issues. Founded by Dr Tara Garnett in 2005, the FCRN 'conducts, synthesises, and communicates research at the intersection of food, climate, and broader sustainability issues' (FCRN, n.d.). Through this work, and in particular the production of a series of reports on sustainable diets, it appears to have a major influence on policy debates in this area.

Table 6.2 Selected reports on sustainable diets

Title	Organisation	Year
Livestock's long shadow: environmental issues and options	Food and Agriculture Organization of the UN	2006
Environmentally effective food choices: proposal notified to the EU, 15.5.2009	National Food Administration and Swedish Environmental Protection Agency	2009
How low can we go? An assessment of greenhouse gas emissions from the UK food system and the scope to reduce them by 2050	Food Climate Research Network, WWF-UK	2009
Setting the table: advice to government on priority elements of sustainable diets	Sustainable Development Commission	2009
Guidelines for a healthy diet: the ecological perspective	Health Council of the Netherlands	2011
2011 double pyramid: healthy food for people, sustainable for the planet	Barilla Center for Food and Nutrition	2011
Sustainable diets and biodiversity: directions and solutions for policy, research and action	Food and Agriculture Organization of the UN, Biodiversity International	2012
On our plate today: healthy, sustainable food choices (Livewell for LIFE final report)	WWF-UK, Friends of Europe	2014
People, plate and planet – the impact of dietary choices on health, greenhouse gas emissions and land use	Centre for Alternative Technology, Wales	2014
What is a healthy and sustainable diet? A discussion paper	Food Climate Research Network	2014
Policies and actions to shift eating patterns: What works?	Food Climate Research Network, Chatham House and EAT Initiative	2015
The principles of healthy and sustainable eating patterns	BBSRC Global Food Security programme	2015
Plates, pyramids and planet – developments in national healthy and sustainable dietary guidelines: a state of play assessment	Food Climate Research Network, Food and Agriculture Organization of the UN	2015
The Eatwell Guide: a more sustainable diet	The Carbon Trust	2016
Sustainable diets for healthy people and a healthy planet	UN System Standing Committee on Nutrition	2016
Eating for 2 degrees: new and updated Livewell plates	World Wide Fund for Nature	2017
Grabbing the bull by the horns: it's time to cut industrial meat and dairy to save the climate	GRAIN	2017

Discussion and next steps

In the period that I was conducting this initial research the topic of sustainable diets, and particularly reduced meat consumption, started to be more widely reported in the British mainstream press. There was widespread reporting of announcements that China and the Netherlands were producing new guidelines advising citizens to reduce their consumption of red meat on environmental grounds (e.g. Gustin, 2016; Millman and Leavenworth, 2016). The case of China is particularly significant, as growth in Chinese meat consumption is thought to be driving global increases and forecast to carry on doing so as its citizens demand levels of consumption equivalent to those of the US and western Europe (Foresight, 2011). In the current phase of this pilot project I have been interviewing expert stakeholders to investigate the ways in which sustainability criteria are being incorporated (or not) into existing dietary guidelines. In exploring these processes further, I aim to understand how different kinds of evidence – from agricultural, environmental and nutrition research – are being used at the national or sub-national levels, and whether the slow and erratic integration between food, environment and health policy is taking place.

I chose to conduct my research in Denmark and Scotland for several reasons. Firstly they are both small northern European countries and members of the EU. Both countries also have large export-oriented livestock industries – pork in the case of Denmark and beef in the case of Scotland. The economic importance of livestock production in both countries makes advice for individuals to reduce their meat consumption potentially very politically contentious. Denmark introduced an unsuccessful 'fat tax' in 2008 (Vallgårda, Holm, and Jensen, 2015) and the UK government has just introduced (in April 2018) a levy on certain sweetened beverages, so both governments have tried to use fiscal methods to reduce consumption of 'unhealthy' products. However, they appear to present contrasting cases with respect to sustainability. Denmark has developed a new and successful model of food policy that some argue Scotland should seek to emulate (Scottish Food Coalition, n.d.). Moreover, Denmark is a welfare state country with one of the world's largest organic market sectors (Wier et al., 2008), which appears to be an important factor in its new approach to food policy. Despite the Scottish government's Nordic ambitions, the UK currently has a much more market-oriented approach to food policy and a smaller organic sector (ibid).

In both countries, the production of official dietary guidelines is a process that spans more than one country rather than being the sole responsibility of the individual nation. Public health nutrition in Scotland is the responsibility of Food Standards Scotland (which is part of the devolved Scottish government) whereas the official dietary guidelines remain the UK-wide Eatwell Guide produced by Public Health England (FAO, 2018).[3] The Eatwell Guide contains two brief mentions of sustainability – it aims to 'help you get a balance of healthier and more sustainable food' and individuals should eat 'two portions of sustainably sourced fish per week' – no further guidance is given on how to achieve this

(PHE, 2016). The official Danish dietary guidelines (or *Kostråd*) are based on the Nordic Nutrition Recommendations (NNR), which are the evidence base for dietary guidelines in Denmark, Finland, Iceland, Norway and Sweden. The NNR is produced by a steering group of researchers from the five countries who synthesise the result from a large number of specially commissioned systematic reviews. The most recent version (Nordic Council of Ministers, 2014) has a standalone chapter on sustainability written by a Swedish nutrition researcher. However, this material is not integrated with the rest of the document, which follows a standard format to report the results of the systematic reviews, and nor is it referred to in the *Kostråd*.

The Livewell Principles, described previously, encapsulate a developing consensus that a sustainable diet is low(er) in meat, fish and dairy (especially red meat because of its high environmental impact); higher in whole-grain cereals, pulses and field-grown fruit and vegetables; and low in high-fat, high-sugar and high-salt processed foods – what are referred to as ultra-processed foods (Monteiro et al., 2017) or discretionary foods (FSS, 2015). Although the environmental benefits of eating local produce is disputed (Edward-Jones et al., 2008), advice to eat locally and seasonally is often included. Based on FAO (2012) estimates that approximately 30 percent of all food produced is spoiled or wasted, advice to reduce the amount of food wasted is also included.

However, there remain significant areas of debate about the application of these principles and I will now outline three key examples that have formed an important part of the stakeholder interviews I have recently undertaken. The first of these is the role of livestock and livestock products in sustainable diets. Animal products, particularly beef and lamb, are highly resource intensive to produce, and livestock farming can result in significant biodiversity loss, land degradation, water stress and pollution. It is argued that in order for diets to become more sustainable, individuals in rich industrialised countries will need to reduce their meat consumption significantly (Reynolds et al., 2014). Eleven of the articles in my collection address this topic directly, mostly by analysing how low levels of meat consumption need to go in order to produce significant reductions in GHG emissions and how well such reductions will be tolerated by the many individuals who appear to value frequent meat consumption. However, in northern Europe, not only are meat and dairy products still important and valued components of diet, but much of the land is unsuitable for cultivation – in Scotland 70 percent of agricultural land falls into this category. If used for food production at all, such land can only be used for grazing sheep and cows. This means that when assessing the impact of livestock products we need to consider the different production systems. Small-scale, extensive grazing systems where animals make use of land that cannot be used for cultivation are not as environmentally damaging as intensive systems where large numbers of animals are fed on crops such as soya and maize that humans could eat (Garnett, 2015). The idea of 'ecological leftovers' (Roos, 2016) where animals are fed only on waste material and foodstuffs that humans cannot or will not eat is one way of developing more resource-efficient livestock production systems.

The second complexity is fish consumption, which is widely acknowledged as both nutritionally valuable and environmentally damaging. Many dietary guidelines recognise the nutritional importance of fish, especially oily fish. For example, the Danish dietary guidelines recommend eating a total of 350 g of fish per week, of which about 200 g should be fish such as salmon, trout, mackerel and herring (Fødevarestyrelsen, 2013). However, global fish stocks are under severe pressure. According to the FAO (2012), 80 percent of the world's fish stocks for which assessment information is available are reported as fully exploited or over-exploited. Moreover, in 2014, around half of Scottish adults (48 percent) ate white fish at least once a week and a quarter (25 percent) ate oily fish, such as mackerel, at least once a week (Scottish Government, 2014). Unless fishing practices and dietary preference change dramatically, doubling consumption will lead to even more unsustainable levels of extractions. GHG emissions do not capture these sorts of environmental impacts, which is perhaps why this issue was rarely mentioned in my collection of articles, apart from in the literature/evidence reviews.

Finally, it may seem as if the relationship between the healthiness of a diet and its sustainability appears to be straightforward: healthier diets contain a lower proportion of animal products and a higher proportion of plant-based foods (which have a lower environmental impact) so the overall diet should, in theory, also have a lower impact. However, some studies (e.g. Masset et al., 2014) show that healthier diets are not necessarily lower in emissions, and in some cases can be higher. This appears to be the result of two factors: (1) the high energy density and low GHG emissions of processed foods such as sweets and salted snacks (Payne, Scarborough, and Cobiac, 2016) and (2) that replacing meat with fruit and vegetables may actually increase the GHG emissions of diets in order to compensate for a reduction in energy density (Masset et al., 2014). There are related debates about what reduction in average meat consumption is necessary to mitigate environmental harms whilst maintaining nutritionally adequate diets (Barré et al., 2018). A final factor influencing the environmental impact of diets high in fruit and vegetables may be the shift away from relatively robust field-grown vegetables, such as cabbage and apples (for the UK), towards delicate greenhouse-grown and/or imported vegetables such as avocados, peppers, baby salad leaves and year-round strawberries (Schoen and Lang, 2016). Again, method of production is important.

Conclusion

In this initial piece of research into sustainable diets, I describe the terrain of this new body of knowledge. It is widely accepted that the concept of sustainability is complex, spans many research areas and can be measured using a range of different metrics. However, my literature review shows much of the research into sustainable diets is published in nutrition journals; and that, within these articles, the environmental impact of the food system is largely assessed in terms of the associated GHG emissions. This means that this new account focuses heavily on the quantifiable elements of sustainable diets – the nutritional balance of different diets and their associated carbon dioxide and methane emissions. When incorporated into policy literature, such highly technical approaches result in strikingly de-politicised accounts of the relationship between food, health and environment. Rather than calling for food system reforms, the resulting policy narratives often put the onus on individuals to mitigate these effects by modifying their eating behaviour.

A more comprehensive understanding of the negative impacts of food production would require collecting and synthesising research evidence from a wide range of areas including (but not limited to) agriculture and agronomy, ecology, energy consumption and nutrition. However, it is not yet clear how these different kinds of evidence can be integrated and what weight should be given to different negative outcomes. This task is made more difficult by the complexity of contemporary food industry supply chains. As the 2013 European horsemeat scandal demonstrated, some areas of food production are not fully understood by governments and food industry actors, let alone properly regulated. This means that accurate data will be hard to collect.

Assessing the environmental impacts of different foods and diets is an urgent and important task that requires the continuing integration of nutrition and sustainability research. However, there is a danger that if the current research trajectory continues, the concept of sustainability as applied to the food system will become narrower, focusing on criteria that are easily quantified at the expense of more integrated and interdisciplinary accounts of contemporary food systems. Hornberg (cited by Friedmann, 2017) argues that managerial approaches – those that work within existing systems rather than attempt radical change – already dominate contemporary sustainability policy. They may also come to dominate debates about how to decrease the environmental impacts of our food choices, marginalising issues such as workers' conditions and animal welfare for which quantitative metrics are less available and appropriate. Such a narrow focus on the measurable aspects of sustainability also seems likely to make dietary guidelines even more remote from current eating practices, and even less relevant to the large majority of individuals and institutions that do not currently engage with them.

Notes

1   Discussions of sustainable diets focus on individual consumption, and are one component of a larger academic and policy literature on sustainable food systems. This literature has also grown significantly in the last ten years, and now has its own journal and textbooks (e.g. Gliessman, 2014).
2   The Mediterranean diet featured in ten articles, many produced by a group of authors affiliated to the FAO office in Rome. They are developing a new identity for the Mediterranean diet as an exemplar of a sustainable diet and being more successful than proponents of the New Nordic Diet, which featured in only two articles.

3 In England the responsibility for dietary guidelines was transferred from the Food Standards Agency to DEFRA in 2010 and then Public Health England in 2013, whereas in
Scotland Food Standards Scotland retains this responsibility.

References

Auestad, N. and Fulgoni, V.L. (2015). What current literature tells us about sustainable
diets: Emerging research linking dietary patterns, environmental sustainability, and economics. *Advances in Nutrition*, 6(1), pp. 19–36.

Barling, D., Lang, T. and Caraher, M. (2002). Joined-up food policy? The trials of governance, public policy and the food system. *Social Policy & Administration*, 36(6),
pp. 556–574.

Barré, T., Perignon, M., Gazan, R., et al. (2018). Integrating nutrient bioavailability and
coproduction links when identifying sustainable diets: How low should we reduce meat
consumption? *PLoS ONE*, 13(2), art. no. e0191767.

Black, J. (2001). Decentring regulation: Understanding the role of regulation and self-
regulation in a 'post-regulatory' world. *Current Legal Problems*, 54(1), pp. 103–146.

Bufton, M.W. (2005). British expert advice on diet and heart disease c.1945–2000. In:
V. Berridge, ed., *Making health policy: Networks in research and policy after 1945*. Amsterdam and New York: Rodopi.

Bufton, M.W. and Berridge, V. (2000). Post-war nutrition science and policy making in Britain c.1945–1994: The case of diet and heart disease. In: D.F. Smith and J. Phillips, eds.,
Food, science, policy and regulation in the twentieth century. London and New York: Routledge.

Carson, R. (1965). *Silent spring*. London: Penguin Books.

DEFRA (2016). *Food statistics pocketbook 2016*. London: Department of Food and
Rural Affairs. Available at: www.gov.uk/government/statistics/food-statistics-pocket-
book-2016 [Accessed 8 Feb. 2019].

Delormier, T., Frolich, K.L. and Potvin, L. (2009). Food and eating as social practice –
understanding eating patterns as social phenomena and implications for public health.
Sociology of Health and Illness, 31(2), pp. 215–228.

Edwards-Jones, G., Milà i Canals, L., Hounsome, N., et al. (2008). Testing the assertion
that 'local food is best': The challenges of an evidence-based approach. *Trends in Food
Science & Technology*, 19(5), pp. 265–274.

FAO (2012). *Sustainable diets and biodiversity: Directions and solutions for policy, research
and action*. Rome: Nutrition and Consumer Protection Division, Food and Agriculture
Organization of the United Nations.

FCRN (n.d.). Food climate research network knowledge for better food systems/about
FCRN (website). Available at: https://fcrn.org.uk/about [Accessed 8 Feb. 2019].

Fødevarestyrelsen [Food Administration] (2013). *De officielle kostråd* [The official dietary
advice]. Copenhagen: Ministry of Food, Agriculture and Fisheries. Available at: http://
altomkost.dk/deofficielleanbefalingertilensundlivsstil/de-officielle-kostraad/ [Accessed
8 Feb. 2019].

Foresight. Tackling Obesities: Future Choices (2007a). *Final project report*. London: The
Government Office for Science.

Foresight. Tackling Obesities: Future Choices (2007b). *Obesogenic environments – evidence
review*. London: The Government Office for Science.

Foresight. The Future of Food and Farming (2011). *Final project report: Executive summary*.
London: The Government Office for Science.

Freidberg, S. (2015). It's complicated: Corporate sustainability and the uneasiness of life cycle assessment. *Science as Culture*, 24(2), pp. 157–182.

Freidberg, S. (2016). Wicked nutrition: The controversial greening of official dietary guidance. *Gastronomica: The Journal of Critical Food Studies*, 16(2), pp. 69–80.

Friedmann, H. (2017). Paradox of transition: Two reports on how to move towards sustainable food systems. *Development and Change*, 48(5), pp. 1210–1226.

FSS (2015). *The Scottish diet: It needs to change.* Aberdeen: Food Standards Scotland. Available at: www.foodstandards.gov.scot/downloads/Final_Report.pdf [Accessed 8 Feb. 2019].

Garnett, T. (2014). *What is a healthy and sustainable diet? (A discussion paper).* Food Climate Research Network. Available at: https://fcrn.org.uk/sites/default/files/fcrn_what_is_a_sustainable_healthy_diet_final_0.pdf [Accessed 8 Feb. 2019].

Garnett, T. (2015). *Gut feelings and possible tomorrows: (where) does animal farming fit?* Food Climate Research Network. Available at: www.fcrn.org.uk/sites/default/files/fcrn_gut_feelings.pdf [Accessed 8 Feb. 2019].

Garrety, K. (1997). Social worlds, actor-networks and controversy: The case of cholesterol, dietary fat and heart disease. *Social Studies of Science*, 27(5), pp. 727–773.

Gieryn, T. (1999). *Cultural boundaries of science: Credibility on the line.* Chicago: University of Chicago Press.

Gliessman, S.R. (2014). *Agroecology: The ecology of sustainable food systems.* 3rd ed. Boca Raton, FL: CRC Press.

Gussow, J.D. (1999). Dietary guidelines for sustainability: Twelve years later. *Journal of Nutrition Education*, 31(4), pp. 194–200.

Gussow, J.D. and Clancy, K.L. (1986). Dietary guidelines for sustainability. *Journal of Nutrition Education and Behaviour*, 18(1), pp. 1–5.

Gustin, G. (2016). *Another nation trims meat from diet advice.* National Geographic, The Plate Food Blog, 23 June 2016. Available at: http://theplate.nationalgeographic.com/2016/03/23/another-nation-trims-meat-from-diet-advice/ [Accessed 8 Feb. 2019].

Hilgartner, S. (2000). *Science on stage: Expert advice as public drama.* Stanford, CA: Stanford University Press.

Jahns, L., Davis-Shaw, W., Lichtenstein, A.H., et al. (2018). The history and future of dietary guidance in America. *Advances in Nutrition*, 9(2), pp. 136–147.

Jasanoff, S. (1990). *The fifth branch: Science advisors as policymakers.* Cambridge, MA: Harvard University Press.

Jasanoff, S. (2005). *Designs on nature: Science and democracy in the Europe and the United States.* Princeton, NJ: Princeton University Press.

Johns, D.M. and Oppenheimer, G.M. (2018). Was there ever really a "sugar conspiracy"? *Science*, 359(6377), pp. 747–750.

Jones, A.D., Hoey, L., Blesh, J., et al. (2016). A systematic review of the measurement of sustainable diets. *Advances in Nutrition*, 7(4), pp. 641–664.

Kates, R. (2011). What kind of a science is sustainability science? *Proceedings of the National Academy of Sciences*, 108(49), pp. 19449–19450.

La Berge, A.F. (2008). How the ideology of low fat conquered America. *Journal of the History of Medicine and Allied Sciences*, 63(2), pp. 139–177.

Lang, T. and Barling, D. (2012). Food security and food sustainability: Reformulating the debate. *The Geographical Journal*, 178(4), pp. 313–326.

Lang, T., Barling, D. and Caraher, M. (2009). *Food policy: Integrating health, environment and society.* Oxford: Oxford University Press.

Lang, T. and Heasman, M. (2004). *Food wars: The global battle for mouths, minds and markets*. London: Earthscan.

Lappé, F.M. (1973). *Diet for a small planet*. New York: Ballantine Books.

Lawrence, F. (2013). Horsemeat scandal: The essential guide. *The Guardian Newspaper*, 15 Feb. 2013, London. Available at: www.theguardian.com/uk/2013/feb/15/horsemeat-scandal-the-essential-guide [Accessed 8 Feb. 2019].

Livewell for LIFE (2014). *Livewell for LIFE final recommendations*. Godalming, UK: WWF-UK and Friends of Europe.

Macdiarmid, J., Kyle, J., Horgan, G., et al. (2011). *Livewell – a balance of healthy and sustainable food choices*. Godalming, UK: WWF-UK.

Mason, P. and Lang, T. (2017). *Sustainable diets: How ecological nutrition can transform consumption and the food system*. London and New York: Earthscan (Routledge).

Masset, G., Vieux, F., Verger, E.O., et al. (2014). Reducing energy intake and energy density for a sustainable diet: A study based on self-selected diets in French adults. *American Journal of Clinical Nutrition*, 99(6), pp. 1460–1469.

Maxwell, S. (1996). Food security: A post-modern perspective. *Food Policy*, 21(2), pp. 155–170.

Millman, O. and Leavenworth, S. (2016). China's plan to cut meat consumption by 50% cheered by climate campaigners. *The Guardian Newspaper*, 20 June 2016, London. Available at: www.theguardian.com/world/2016/jun/20/chinas-meat-consumption-climate-change [Accessed 8 Feb. 2019].

Monteiro, C., Cannon, G., Moubarac, J., et al. (2017). The UN decade of nutrition, the NOVA food classification and the trouble with ultra-processing. *Public Health Nutrition*, 21(1), pp. 5–17.

Nestle, M. (2002). *Food politics: How the food industry influences nutrition and health*. Berkley, LA and London: University of California Press.

Nordic Council of Ministers. (2014). *Nordic nutrition recommendations 2012: Integrating nutrition and physical activity*. Copenhagen: Nordic Council of Ministers. Available at: https://norden.diva-portal.org/smash/get/diva2:704251/FULLTEXT01.pdf [Accessed 8 Feb. 2019].

Payne, C.L., Scarborough, P. and Cobiac, L. (2016). Do low-carbon-emission diets lead to higher nutritional quality and positive health outcomes? A systematic review of the literature. *Public Health Nutrition*, 19(14), pp. 2654–2661.

Popkin, B.M. (2004). The nutrition transition: An overview of world patterns of change. *Nutrition Reviews*, 62, pp. S140–S143.

Public Health England (2016). *The Eatwell guide*. London: Department of Health. Available at: www.gov.uk/government/publications/the-eatwell-guide [Accessed 8 Feb. 2019].

Reynolds, C.J., Buckley, J.D., Weinstein, P., et al. (2014). Are the dietary guidelines for meat, fat, fruit and vegetable consumption appropriate for environmental sustainability? A review of the literature. *Nutrients*, 6(6), pp. 2251–2265.

Ridoutt, B.G., Hendrie, G.A. and Noakes, M. (2017). Dietary strategies to reduce environmental impact: A critical review of the evidence. *Advances in Nutrition*, 8(6), pp. 933–946.

Roos, E. (2015). *Environmental concerns now in Sweden's newly launched dietary guidelines*. FCRN Blog, 11 June 2015. Available at: www.fcrn.org.uk/fcrn-blogs/elin-roos/environmental-concerns-now-sweden%E2%80%99s-newly-launched-dietary-guidelines [Accessed 8 Feb. 2019].

Schoen, V. and Lang, T. (2016). *Horticulture in the UK: Potential for meeting dietary guideline demands.* Food Research Collaboration Policy Brief. Available at: https://foodresearch.org.uk/publications/horticulture-in-the-uk/ [Accessed 8 Feb. 2019].

Scottish Food Coalition (n.d.). *Events/food: A solution to a health crisis. Taking inspiration from the Nordic approach* (webpage). Available at: www.foodcoalition.scot/food-a-solution-to-a-health-crisis.html [Accessed 8 Feb. 2019].

Scottish Government. (2014). *The Scottish health survey 2014: Volume 1: Main report.* Available at: www.gov.scot/Publications/2015/09/6648/318775 [Accessed 8 Feb. 2019].

Shaw, J.D. (2008). *World food security: A history since 1945.* Basingstoke and New York: Palgrave Macmillan.

Smith, D.F and Bufton, M.W. (2004). A case of Parturiunt Montes, Nascetur Ridiculus Mus? The BMA nutrition committee 1947–50 and the political disengagement of nutrition science. *Journal of the History of Medicine and Allied Sciences,* 59, pp. 240–272.

The Lancet. (2016). Life, death, and disability in 2016 (Editorial). *The Lancet,* 390(10100), p. 1083.

Tyszler, M., Kramer, G. and Blonk, H. (2016). Just eating healthier is not enough: Studying the environmental impact of different diet scenarios for Dutch women (31–50 years old) by linear programming. *International Journal of Life Cycle Assessment,* 21(5), pp. 701–709.

UN Department of Public Information (2010). *Resumed review conference on the agreement relating to the conservation and management of straddling fish stocks and highly migratory fish stock.* Available at: www.un.org/depts/los/convention_agreements/reviewconf/FishStocks_EN_A.pdf [Accessed 8 Feb. 2019].

UN World Commission on Environment and Development (1987). *Our common future.* Oxford: Oxford University Press.

Vallgårda, S., Holm, L. and Jensen, J.D. (2015). The Danish tax on saturated fat: Why it did not survive. *European Journal of Clinical Nutrition,* 69, pp. 223–226.

van de Kamp, M.E., van Dooren, C., Hollander, A., et al. (2018). Healthy diets with reduced environmental impact? – The greenhouse gas emissions of various diets adhering to the Dutch food based dietary guidelines. *Food Research International,* 104, pp. 14–24.

Vetőné Mózner, Z. (2014). Sustainability and consumption structure: Environmental impacts of food consumption clusters. A case study for Hungary. *International Journal of Consumer Studies,* 38(5), pp. 529–539.

Walker, C. and Cannon, G. (1985). *The food scandal.* London: Century Publishing.

WHO (1990). *Diet, nutrition, and the prevention of chronic diseases.* Geneva: World Health Organization (Technical Report Series 797).

WHO Europe Office (2014). *European food and nutrition action plan 2015–2020.* Copenhagen: WHO Regional Office for Europe.

Wier, M., O'Doherty Jensen, K., Mørch Andersen, L., et al. (2008). The character of demand in mature organic food markets: Great Britain and Denmark compared. *Food Policy,* 33(5), pp. 406–421.

WWF-UK (2017). *Eating for 2 degrees – new and updated livewell plates (summary report).* Available at: www.wwf.org.uk/eatingfor2degrees [Accessed 8 Feb. 2019].

Wynne, B. (1996). May the sheep safely graze? In: S. Lash, B. Szersynski and B. Wynne, eds., *Risk, environment and modernity: Towards a new ecology.* London: SAGE Publications.

Yearley, S. (2000). Making systematic sense of public discontents with expert knowledge: Two analytical approaches and a case study. *Public Understanding of Science,* 9(2), pp. 105–122.

Chapter 7

Healthy eating, social class and ethnicity

Exploring the food practices of South Asian mothers

Punita Chowbey

Introduction

The diets of South Asian populations are of particular concern due to poorer health outcomes associated with the consumption of traditional food in the UK (Lawton et al., 2008; Ludwig, Cox, and Ellahi, 2011; Qureshi, 2019). However, this concern has not been translated into a high-quality evidence base. Most research to date has focused on healthy eating in the context of health issues such as diabetes and growing obesity resulting from the dietary habits of South Asians (for example, Lawton et al., 2008; Pallan, Parry, and Adab, 2012). Much of this literature has examined diets and physical activities side by side and has often focused on the type and composition of food consumed by South Asians, for example the consumption of saturated fat and fruits and vegetables (Sevak et al., 2004; Anderson et al., 2005; Ludwig, Cox, and Ellahi, 2011; Lawton et al., 2008).

Some studies have attempted to understand the meaning of healthy eating and rationale for food choice among South Asians. The majority of these studies show that South Asians tend to report some aspects of 'Asian food' such as spice and use of oil as being bad for health and at the same time the use of traditional system based on food being digestible/indigestible, hot/cold, strong/weak are considered healthy (Bradby, 1997; Chowdhury, Helman, and Greenhalgh, 2000; Qureshi, 2019; Jamal, 1998). Qureshi (2019) in her study of long-term health conditions among Pakistanis shows that her respondents had internalised the idea of 'Asian food' being bad, with ill-health resulting from the fat, spices and sugar of the 'Asian diet', often under the influence of health providers' advice. However, they also sometimes positively judged 'home food' compared with 'English food', which was taken as a synonym for junk. Bradby's (1997) study in Glasgow revealed women's understanding of healthy eating in terms of a combination of medical orthodoxy of healthy eating and alternative beliefs of food rooted in the folk beliefs and advice of elders. Some studies suggested resistance against stereotypes, such as Asian food being only curry, and respondents reporting eating from a variety of cuisine (Wyke and Landman, 1997).

Researchers have also attempted to study influences on healthy eating in the British South Asian population (Chowbey and Harrop, 2016; Pallan, Parry, and Adab, 2012; Anderson et al., 2005; Lawrence et al., 2007). For example, Pallan, Parry, and Adab (2012) examine influences on the development of obesity in British South Asian children and have identified a range of influences including individual, family and cultural influences as well as influences of school and local and macro-environment. Chowdhury, Helman and Greenhalgh (2000) identified structural and economic factors such as affordability and availability as well as cultural influences on British Bangladeshi's food beliefs and practices. They identified religious restrictions on food items as an important factor in food practices. Emadian et al. (2017) have identified cultural commitments including faith events, motivation and time as key barriers to eating healthily in their research with obese South Asian men living in the UK. They identify family support as an important facilitator to dietary change. There has also been an increased interest in the role of migration history on food practices where research has suggested more fat and sugar is consumed following migration (Williams et al., 1998; Chowdhury, Helman, and Greenhalgh, 2000; Anderson et al., 2005). Further, changes in food habits over generations have also been noted, with fewer traditional meals amongst members of the so-called second-generation than among first-generation migrants (Gilbert and Khokhar, 2008; Lofink, 2012). However, such research overlooks how transnational marriage and caregiving practices affect household food practices, such as when new migrant family members join households. A few studies have noted diversity in food practices among the various South Asian ethno-religious groups. For example, a preference for English-style breakfasts consisting of bread and cereals was found to be more common among Bangladeshi and Pakistani than Hindus and Sikhs, who preferred a cooked breakfast of chapattis and parathas (Gilbert and Khokhar, 2008). In spite of this recognition of diversity, however, and perhaps surprisingly, the influence of social class on South Asian food practices has been neglected in the literature. There does not appear to be any study published to date which examines South Asian food practices in relation to socioeconomic differentiation within the population. Class-based distinctions in food preferences, which have been studied in the white British population, are often conflated with ethnicity-based preferences. Social class, which 'acts as a structural determinant shaping access to food, and especially to food that is healthy, appealing and desired' (Smith Maguire, 2016, p. 12), is an important dimension currently missing from the literature on food practices among South Asian population.

Further, employment has been shown to have an impact on household food practices in terms of the consumption of home-cooked food and types of food prepared at home (Blake et al., 2011; Devine et al., 2006, 2003; Roos et al., 2007). These studies suggest that households from low-income backgrounds, employed in non-standard working hours or single parenthood are likely to eat less home-cooked meals and rely on unhealthy frozen and chilled convenience food. For example, Blake et al. (2011) found that married men with non-employed spouses had more home-cooked meals. However, the majority of these studies examining the impact of income and time on household shopping, cooking and eating practices has been done with the white British families or in western countries. Research on South Asian women's experiences of food practices in context of motherhood and (re-)entry to labour market is scant. An exception to this is Ellen's (2017) work with British South Asian women, in which she has explored their ability to negotiate new gender identities as both good mothers and independent women through mothering and consumption practices including food.

South Asians comprise 4.9 percent of the total UK population (2011, UK Census). Indians (including the Gujarati Hindus who were the focus of my study) and Pakistanis are the two largest ethnic groups. Gujarati Hindus and Pakistani Muslims are culturally similar in several respects, such as transnational practices in the areas of marriage, caring and financial obligations. At the same time, significant differences exist in their socioeconomic positions and migration histories. Women's employment differs across both groups. Pakistani women (aged 25–49 years) have relatively lower employment rates with 43 percent being economically active and with an unemployment rate three times that of white British women in the UK (Nazroo and Kapadia, 2013). For Indian groups, including Gujarati Hindus, the economic activity rate for women aged 25–49 years was 79 percent (Nazroo and Kapadia, 2013). Evidence suggests that being a mother of young children and educational qualifications have more impact than their migration history (Dale and Ahmed, 2011). Having young children further compromises women's position in the labour market, leading sometimes to withdrawal from work, underemployment or significant effort to maintain participation in the context of gender and ethnic labour market discrimination. The existing research shows demands of work having impact on food choices and the need to balance family values such as closeness and personal achievement (Devine et al., 2003). However such evidence is negligible in the context of intersection of ethnicity and social class.

Against this background, the chapter explores the narratives of mothers from diverse socioeconomic and ethnic backgrounds regarding food practices in the context of healthy eating. These mothers are both first- and second-generation migrants with dependent children from two South Asian groups in Britain: Pakistani Muslims and Gujarati Hindus. It will explore women's understanding of eating healthily, the practices they engage in and their experiences of healthy eating interventions and messages within diverse socioeconomic locations. The chapter will also examine how social class, as indicated by their occupation and income, influences healthy eating narratives, practices and aspirations among this sample. It challenges class-neutral policies and practices pertaining to healthy eating and the assumption that South Asian food practices are homogeneous, static and unhealthy and that healthy eating messages and initiatives will therefore engender transformative change.

Understanding food practices and social class among South Asians

To analyse food practices based on social class, I draw upon Bourdieu's (1984, 1987) exposition of class as a structured and structuring micro-practice to understand how people's differing access to social, cultural, economic and symbolic capital translates into food practices embedded in socioeconomic and cultural locations. For example, food-related cultural capital (in an *'embodied'*, *'objectified'* and *'institutionalised'* state, Bourdieu, 2011, p. 47) includes visits to restaurants, knowledge of various aspects of healthy eating and references to research and possession of books, especially on food. I also consider respondents' social capital through spontaneous references made to eating and hanging out with friends, colleagues and relatives. Symbolic capital, 'which is the form the different types of capital take once they are perceived and recognised as legitimate' (Bourdieu, 1987, p. 4), is manifested as the legitimacy, honour or prestige attached to food choice and practices.

Bourdieu (1984, p. 166) argues that '[t]he habitus is necessity internalized and converted into a disposition that generates meaningful practices and meaning – giving perceptions; it is a general, transposable disposition which carries out a systematic, universal application – beyond the limits of what has been directly learnt-of the necessity inherent in the learning conditions'. He maintains that our practices and behaviours are a result of habituated ways of thinking, talking and behaving (our *habitus*) which we take as given. Bourdieu (1984, p. 173) distinguishes between 'substance' and 'form' and identifies taste as 'the real principle of preference' which is shaped very early on in life. He argues that taste is a forced choice as it exists in the absence of alternatives. Bourdieu's theorisation of various forms of capitals and *habitus* challenges the issues of lifestyle simply in terms of personal choice (Williams, 1995). Williams argues that Bourdieu's analysis is helpful in understanding the issues of class, lifestyle and health and argues for 'an approach which recognises the dialectical interplay of freedom and constraint in daily life and accords equal weight to both elements' (Williams, 1995, p. 601).

However, recent research has begun to challenge the relevance of the supposed divide between middle-class discernment and working-class necessity in food practices while maintaining the relevance of class dynamics (Smith Maguire, 2016; Beagan, Chapman, and Power, 2016; Flemmen, Hjellbrekke, and Jarness, 2018). Smith Maguire has argued to look beyond the affordability and necessity in working class food practices and revisit 'Bourdieu's (1984) conceptualization of working class habitus and taste, and the notion of a "taste of necessity". There is a need to develop a more nuanced, dynamic account of the tastes of low-income and economically marginal groups' (2016, p. 16).

Moreover, given the cumulative effects of migration history and minority status, it is difficult to apply the existing class discourses denoted by social, human, cultural and symbolic capital to South Asians in the UK. Research has highlighted the complexities involved in identifying ethnicity with a specific social class (Song, 2003; Modood, 2004; Daye, 2016; Maylor and Williams, 2011; Rollock et al., 2013). Rollock et al. attribute some of these complexities to 'the relative newness of the Black middle classes and respondents' broadly similar working-class trajectories alongside ongoing experiences of racism within a society that privileges and gives legitimacy to a dominant White middle-class norm' (2013, p. 262). However, a class-based analysis remains relevant in the context of ethnicity because it resonates with the lived experiences of individuals and families (Blake et al., 2011; Archer, 2011).

Methods

The aim of the study was to explore the relationship between access to and control over resources in household food practices amongst women from diverse migration backgrounds in Britain, India and Pakistan. This chapter presents findings from fieldwork conducted in Britain which was conducted in five cities in England (Sheffield, Rotherham, Bradford, London and Nottingham). The research was informed by an intersectional paradigm 'that establishes that social existence is never singular, but rather that everybody belongs simultaneously to multiple categories that are historically and geographically located and that shift over time' (Phoenix, 2006, p. 28). The research design recognises the differences and similarities in women's experiences based on their ethnicity, nationality, socioeconomic status, occupation, household composition (joint and nuclear), education and migration histories. The constructivist grounded theory approach (Charmaz, 2006) was employed as it lends itself to interpreting complex phenomena. This approach provides 'systematic, yet flexible guidelines for collecting and analysing qualitative data to construct theories "grounded" in data themselves' (Charmaz, 2006, p. 2). It therefore enabled connection between women's experiences of household food practices and the broader socioeconomic context shaping broader healthy eating discourses and subsequent interventions.

Data collection took place from 2013–16. The initial fieldwork included visits to community organisations, attending group sessions on cooking and eating, and going to religious celebrations in temples and community organisations. This was followed by three focus groups that included a participatory exercise involving a matrix of needs, opportunities and constraints to understand household resource allocations and food practices pertaining to eating healthily. These included preferences, knowledge about food and supplies of food. These were followed by three interviews with community leaders to inform the areas to be explored in the interviews. A total of 35 interviews were conducted with first- and second-generation mothers from South Asian backgrounds with dependent children (Pakistani Muslims and Gujarati Hindus) in their preferred languages (Hindi, English and Urdu). These women were recruited through community networks by word of mouth and through visits to various groups. As a mother of a dependent child who is from a South Asian background, I did not find it difficult to access respondents and was able to build a rapport that led to lengthy interviews lasting between 1.5 to 4 hours. However, my familiarity with their experiences also required me to be reflexive and aware of my own beliefs and experiences in constructing rather than discovering research. All the interviews took place at respondents' homes, except for one that took place in a car and two at a community organisation. Data was analysed according to the grounded theory method which involved at least two phases of coding: initial coding which identified 'fragments of data', and focused coding which identified the most useful initial codes (Charmaz, 2006, p. 42). This stage was followed by theoretical coding, which specifies possible relationships between coding categories (Charmaz, 2006). Ethical approval for this research was granted by the Sheffield Hallam University Ethics Committee.

Findings

To address the complexities of a class-informed analysis of South Asian mothers' practices, views and experiences of healthy eating, I utilised the following occupational categories as a proxy for mothers' income: homemaker, both high income and low income; professional/managerial; skilled/clerical/assistant; and manual employment. All manual workers at the time of the research, except one Gujarati mother, had given up work and were looking after children; these were included in the category of low-income homemakers. High-income homemakers were often highly qualified; several held professional positions in the past and were married to professionals or successful businessmen. Those in the professional category may also be married to someone in manual employment and live in a poor neighbourhood. Further, some respondents were underemployed at the time of the research; they were educated and had professional employment in their home country but had to accept a low-skilled job following their migration. These categories are therefore complex and fluid; however, they are meaningful as they resonated with respondents' lived experiences and had an impact on their economic and social capital, the time available for cooking and their ability to access information. The findings are presented in three sections: meaning of healthy eating understood by the respondents in terms of their symbolic and social value instead of nutritional values, healthy eating practices in the context of resources (time and money) and response to healthy eating initiatives.

Meaning of healthy eating: symbolic and social value

Healthy eating was generally understood, across all occupations, as consuming more fruit and vegetables and less oil. Overall, there were few mentions of portion sizes or five food categories, although there were some differences in this area as mothers from professional and some from skilled employment backgrounds made more references to these categories. In comparison, skilled mothers and homemakers from low-income backgrounds tended to focus more on the food itself and volunteered fewer details about eating healthily. Their narratives focused on eating fruit and vegetables or eating certain types of food, Asian or 'English'. For example, Gazala, a first-generation Pakistani homemaker with no formal education who is married to a professional and lives in a poor neighbourhood, offered the following opinion of healthy eating:

> I want to say everything is available in [UK] that is normally good and healthy foods, sometimes it depends upon our choice what we are going to cook for ourselves, so we don't see any problem in here . . . you see my health is very good. But there is nothing like specific for healthy food.

Professional mothers' accounts of healthy eating were more expansive and included not only fruits, vegetables and less fatty products but also the quality of the food items, references to organic food and the use of preservatives and additives. Professional mothers and some skilled workers provided details about food expectations, rules around food and aspirations for healthiness and variety in food practices. Food expectations varied for mothers and included meal timings, presentation, portion, manners and cuisine. Their narratives placed value on what food consumption symbolises in their households – ideals of equality, egalitarian gender roles, intimacy between couples and parental ideologies. Some mothers aspired to present a picture of a family united in its food choice and practices; a family that is egalitarian through the rejection of women serving hot *rotis* (chapattis) and eating only at the end. Professionals who were married to other professionals and high-income homemakers were more likely to present healthy eating practices in unitary terms. Their narratives showed an aspiration to be united as a family regarding their understanding and practice of healthy eating. A second-generation British Gujarati professional, Neha, talked at length about herself and her husband's goals of eating healthily, and of going to the gym and using cooking methods to facilitate this:

> I've got a grill that we use occasionally, but with the actifryer it's really a lot healthier. And I'm not really into eating like samosas and things every day . . . but now that we're both in our forties, we don't really think it's a good idea . . . and I'm a member at the gym, as he is, but he's a lot better at going to it.

However, the small number of professionals married to those in manual employment reported differences in food-related expectations. Hora, a second-generation British Pakistani professional described the culture shock she experienced when she moved to her husband's house following her marriage. She described the dinner table as a democratic space where everyone was equal and free to say or behave as they wished. However, her husband, who left school after GCSEs and started a restaurant business, had grown up in a very different food environment. She stated that:

So, for me I think that [mother-in-law scolding her for wanting chicken she was cooking for her son] set the tone of food expectations. And I think because I had never ever had to deal with something like this before. I grew up where it was more to do, there was so much equality that yes if there was suddenly chicken curry on the table whoever got a piece first got their favourite piece first. Nobody took out a piece like 'oh that's for my son', or 'that's for my daughter' or anybody. But this is what I noticed was happening here. And I felt there was priorities given over the best cut of meat. It was unusual, it was very strange.

Hora linked her mother's eating practices with education and gender equality. For Hora, food communicated equality, sharing and togetherness; however, for Veena, a first-generation British Gujarati professional and daughter of a prominent bureaucrat in India married to another professional, how one ate was symbolic of a proper upbringing and a medium through which to communicate values and spend time together. She discussed the rules about eating in her maternal home at length. She described food time as having a 'set of rules' everyone adhered to and a 'proper' way to eat healthy food that was nourishing and ethical.

Like in my parents' household, there are a few set rules about food. One was that at least one meal, which normally is the dinner, the whole family will have together. And the second was whatever was served, whatever is at the table, you either ate that or you did not.

The narratives of professional and high-income homemakers tended to be more varied than low-income homemakers and skilled mothers. Specifically, there were notable differences across occupations in the language, depth and content of healthy eating discourses. Professional and some skilled mothers often described encouraging their children to try new food, particularly international food, so that they were familiar with different cuisines and could eat in different countries with confidence. They often used negation to express what they do or do not consider to be healthy practices to differentiate themselves from others. For instance, Nasreen, a second-generation British Pakistani professional, commented on her family's stance towards giving fizzy drinks to the children: *'We don't have fizzy, we don't buy fizzy. They've never had, I never stock fizzy drinks at home. If for some reason there is fizzy drink in the home, they won't touch it without asking'.*

These references were often made with respect to the practices and food they considered unhealthy or sometimes traditional food such as curry. For some mothers, curry symbolised not only traditional ways of eating but also labour and a lack of taste for healthy, international and English food. Faridah, a second-generation Pakistani homemaker and a psychology graduate from a prestigious university, distanced her household food practices from those of her relatives and friends who like curry and chapattis: *'He likes both. Yeah. Because there's some men I know like, they want to have their curries and chapattis every single day or at least once a day at least, whereas he's not'.*

Professional mothers and some high-income homemakers often referred to discipline and self-regulation. Similarly, low-income homemakers and some skilled workers made frequent reference to the need for discipline and to teach children self-control. Several professional mothers mentioned having a snack drawer/cupboard/container accessible to the children who were only allowed to take snacks at fixed times, often after school. For instance, a second-generation British Gujarati professional, Rohini, who is married to a second-generation British professional, described a fixed snack drawer time of 4:00 p.m. for her children:

> *Treats, chocolates and stuff, yeah, we have a snack drawer and then the children can have something, limited something, from the snack drawer at four o'clock when they come from school.*

In terms of the social context of healthy eating, most mothers from both ethnic groups and all occupational categories reported eating socially at home with friends and family more often than going out for drinks or meals. References to eating out in McDonald's and Burger King were made by several mothers across all four occupational categories. However, there were some differences based on occupation. For example, some professionals described eating out with other professional friends across different ethnicities and trying new food. Hora describes her experience of eating out as follows:

> *And so I used to go to restaurants, because a lot of socialising was done around food for me and my friends. And we would actually try out new restaurants. London is so huge, you hear about something and even if it's like 20 miles away you will get the tube and get there if you know the food is good. And I wasn't a food snob. So I could happily eat like in a five, like a Michelin star restaurant, but I would happily go to the café, you know, where there's like an auntie at the back who cooks the food, because for me it was to do with taste.*

Whereas Hora stressed an appreciation for taste, several low-income homemakers and some skilled mothers described social eating at home with their friends and relatives, which often involved traditional food. For instance, Johara, a second-generation Pakistani mother in skilled employment married to a first-generation Pakistani migrant in the taxi trade, explained eating at home and the diverse strategies employed for different social networks as follows:

> *It depends who is coming over. Some of my, like, if they're older relatives or anything and they're used to the traditional chapattis and curries and stuff so then I'll make something like that, but if it's my friends or something like that I'll do like a quiche or a lasagne.*

Overall, although there were diverse responses in each occupational category, there appeared to be significant differences in the way they discussed healthy eating, the symbolic and social meaning of food practices and aspirations to eat healthily. In the next section, I examine how occupational and income-related differences translate into healthy eating practices at home.

Healthy eating practices in the context of resources: time and money

In contrast to narratives of healthy eating, there were no clear occupational differences in mothers' reports regarding whether they cook and eat healthily. Both ethnic groups often reported eating healthily and all four occupational categories referred to not being able to eat healthily, although their reasons varied. This contrasts with the diversity in knowledge of healthy eating displayed by professionals and high-income homemakers compared to those in skilled work and low-income homemakers. There was no clear pattern based on occupation in terms of what mothers from different backgrounds cooked on a day to day basis, although low-income homemakers and some skilled workers from a Pakistani background reported cooking *roti* and *salan* more often than professional mothers. Gujarati mothers often referred to cooking vegetarian meals, less use of meat and aspirations for a variety of food per meal. Mothers from a Pakistani Muslim background often referred to having meat or chicken in their diet. Mothers across all occupations referred not only to cooking traditional food but also regularly cooking 'English' food such as pizza, pasta, quiche, jacket potatoes, fish and chips, burgers and sandwiches. However, professionals and high-income homemakers were more likely to differentiate between healthy and unhealthy western food than low-income homemakers and some skilled workers. Several mothers from all four categories reported cooking 'English'/international food in combination with Asian food for the same meal or on different days of the week. For instance, Faridah, a second-generation British Pakistani mother with a university degree, recalled what she had cooked for the last three days:

> Shepherd's pie yesterday with salad, and the day before what did we have? I made chicken palak [spinach] and we had some chapattis and stuff with it, and day before that I think – what did I make? We had some kebabs, mincemeat ones with sandwich and salad and stuff.

A first-generation Gujarati mother in manual work also reported that:

> I try and give my kids . . . I try and give them and give them [South Asian food] at least 4 days a week . . . our food like sabji, roti . . . you know. On the other three days they might have chips or may be jacket potatoes . . . sandwiches.

Although mothers from all classes and both ethnic groups reported cooking from scratch and using fresh ingredients, this was more commonly reported by homemakers. They expressed their dislike for convenience food, although references to the occasional take away, burger or pizza cooked in the oven were not uncommon. However, most strived for home-cooked hot food and nutritious meals for family members.

Some mothers from all occupational backgrounds who were married to men with lower educational qualifications and occupational status reported dissent in their food practices. In many cases, such practices appeared to be influenced by the husbands' preferences, although mothers often carved out a parallel subpractice for their own food and that of their children. In the following quotation, a second-generation British Pakistani mother, Amira, a skilled and university-educated worker married to a first-generation Pakistani in manual employment, talks about how she prepares two different types of meals due to a divergent understanding of healthy food and associated food choices:

> Because he likes his curry and chapatti a lot, so probably I'll be cooking that and may be a few times in the week for him, but me and my son we can eat anything, so . . . I'm like preparing two meals . . . we tend to have English food.

Instead of cooking two separate meals, Bilquis, a second-generation British Pakistani low-income homemaker married to a first-generation Pakistani in semi-skilled employment, described how her food expectations and aspirations to eat healthily have now changed, although she did not offer a clear explanation as to why. She talked about her husband's preference for *salan* and *roti* and how she had now come to like it as well:

> Mainly I used to have like sandwiches or I used to have, I don't know, fish and chips, pilau, grilled fish with vegetables. I used to mainly I preferred it on the English side. Now I've come to the greasy side, salan and roti!

Reports by professional mothers and some high-income homemakers on cooking and feeding children healthily was often made in the context of wider issues linked to emotional, physical and career outcomes for their children and themselves. Compared to mothers from other occupations, their narratives focused less on food per se and included physical activities, weight loss goals and time for self and children. A second-generation British Pakistani, professional mother Nasreen, explained how food is one of the many aspects of caring for her children's needs. She emphasised how spending time helping them to read was an important part of her evening:

> Today I know that I'm working so yesterday I cooked for them, so I know that when they get back from school the food's cooked, that's the main thing. . . . I know that I'll feed him, then if he needs a bath I'll give him a bath, brush his teeth. By 6.30/6.45 I'll say come on, you choose a couple of books and we'll go upstairs. I'll read to him and then he'll fall asleep, and I have to sit next to him while he falls asleep. And then he'll fall asleep eventually.

In contrast, most of the skilled workers and homemakers often reported prioritising fresh cooking over other things, despite time being a constraint. Jahanara, a second-generation Pakistani mother employed in a secretarial position married to a first-generation Pakistani cab driver, stated:

> No, no, I don't [spend a huge amount of time cooking] but it adds up, you know, just feeding them. But by the time I'm finished and they have finished I'm exhausted and they're exhausted. So some days I miss, I forget to think about the homework, and then when we're just relaxing, and then we're just so tired, do you know what I mean?

Thus, while Nasreen reported being organised and feeding her children food cooked a day earlier so that she could spend time with them, Jahanara always cooked fresh meals for her children. Prioritising fresh food leaves little time to help her children with their homework. In some professional households, mothers were able to feed children early and spend time with them because eating and bedtimes were separate for children, as in Nasreen's home. Conversely, in Jahanara's family the rules were more relaxed and children and adults ate when the meal was ready.

As a sub-group, mothers in demanding professions such as consultants working in hospitals and senior academics often reported not being able to eat healthily due to time constraints. Some mothers in highly paid employment talked of cooking as something that had to be fitted in around their work. Kishwar, a first-generation Pakistani professional in a very demanding job, was often unable to find time to cook food for her children:

> I have only recently started making chapatti, otherwise, we used to have Naan [shop bought] or pita bread. Because, my son loves chapatti, so, I have started kneading flour at home, however the chapatti turns out, I am getting better at it. So, I make these things, when I am off, or when I am feeling energetic enough to do it, around the weekend and stuff.

Although working mothers from all occupational backgrounds talked about the things that they could do to eat healthily but were not able to due to work commitments, professional mothers expressed this viewed frequently and sometimes unapologetically. Muskan, a first-generation Pakistani professional, expressed her inability to spend time in the kitchen as follows:

> Last week has been a disaster, so I'm not sure if that's a good week to talk about . . . but generally, yeah we would prefer that we eat food cooked at home, but then what affects is what work you are doing, how much time do you have, that kind of thing . . . if I am at work, when I come back at 6.30–7 in the evening, so then do I have the capacity to stand and cook, that tends to affect a bit.

Muskan further suggests the need for her to work longer hours than others to be accepted and respected due to being from minority ethnic background. Whereas professional mothers often cited time as a major factor in not being able to eat healthily, for mothers in other categories the reasons were more varied and included time, the influences of other family members and budget. Compared to professional mothers, most homemakers from all income backgrounds reported prioritising cooking fresh food, sometimes at the expense of other chores or children's homework. However, women from both ethnic groups and across all occupations resented time spent on cooking and shared strategies to reduce the drudgery involved in preparing hot meals for the family and, in many cases, the extended family. Strategies involved cooking a big pot that lasts several days, creating several meals from one base item (for example, frying chicken for chicken curry, wraps and with salad) and freezing half-prepared food items. Johara explained her reasons for these strategies: *I'll only want to do one thing cooking. I don't want to be stood there all day cooking in the kitchen and stuff.*

A cursory glance suggests there are no clear occupational differences based on budget when eating healthily. Some mothers cited budget as a constraint to eating healthily whereas others felt that, although healthy food can be more expensive, there are cheaper ways of eating healthily and cost should not be a constraint. For instance, Fari, a first-generation Pakistani high-income homemaker who was in professional employment prior to having children and is married to a professional, reported:

> No, I think in any kind of budget you can have things; it depends how you manage things. You can go out and spend a fiver on one small two people lasagne, and then if you buy from that fiver, you know, just the meat, minced meat, that's 1kg of minced meat for £5.50, and then you make your own dough and that costs you, what, 20p of butter and things like that, and then you can make more out of it.

Although some mothers from all class backgrounds felt eating healthily was more expensive, they also stated there were creative ways of doing so on a budget. However, compared to professionals and high-income homemakers, the narratives of low-income homemakers and skilled mothers contained more references to the cost of healthy eating. Their narratives also contained reasons for their food practices – for instance, it is cheaper to eat in an 'Asian way' on a budget as this makes food last longer, and it is economical to cook one big pot for many people. Soha, a second-generation British Pakistani low-income homemaker married to a first-generation Pakistani manual worker, reported:

> Healthy food is a waste of money as well isn't it? How? Tuna is healthy, it's only going to fill you up two sandwiches, yeah, and how expensive are three packets of tuna, £4? How many sandwiches can you make? How many big families can you have? I've got six people in my property living. If it's £4 I buy £2 or £3 chicken,

make a curry and fill everybody up for two days, add a few potatoes in it. I show you a 60p curry, Lehsuni curry.

Amira also reported that she only bought vegetables when they are on offer because of the cost involved, otherwise she buys from a man who provides reasonably priced fruits and vegetables:

Healthy eating is expensive. . . . I mean sometimes, I tend to buy fruits and vegetables [from superstores] if there is a deal . . . but it is expensive, we have, the vegetable and fruit van comes every Thursday, and he is quite reasonable. . . , so I tend to buy my fruit and vegetables from him.

Many manual workers and low-income homemaker mothers spoke of several strategies they adopted to feed their family healthy meals, such as shopping around for cheaper food items, buying items on offer, cooking in bulk and choosing dishes that are nutritious but filling and cheap.

Perhaps surprisingly, there was no clear pattern regarding healthy cooking and eating across the four occupational categories. All categories contained mothers who cooked and ate healthily and mothers who did not. Although time was a constraint for all mothers, low-income homemakers and some skilled workers were able to prioritise shopping and cooking healthy meals for their family despite budgetary constraints although for a variety of reasons, which included cheaper ways to feeding family members and demands of husbands and in-laws as well as the influence of social networks. Conversely, some high-earning professional mothers more often reported not being able to eat as healthily as they wished to due to long working hours and other commitments. Overall, low-income homemakers therefore reported eating more healthily than some professional high-earning mothers.

The reasons for not eating healthily varied and included resources including time and money discussed here, but also other factors such as relationship dynamics and domestic and economic abuse (Chowbey, 2016, 2017), which I have discussed elsewhere. In addition to time and budget, the knowledge and skills involved in cooking and eating healthily were regularly mentioned. These will be examined in the following section in the context of experiences of healthy eating initiatives.

Response to healthy eating initiatives

Less than half of the mothers had received information from formal sources such as healthy eating courses at the community centre, children's schools or from health visitors and GPs. Professional and skilled workers were more likely to cite the internet as a source of information than those in manual work or low-income homemakers. However, a minority of mothers from all occupational backgrounds had engaged with some form of formal education/information on healthy eating.

Among those who had access to these services, most either found the courses to be irrelevant or to have little impact on their day-to-day food practices. Some mothers felt their traditional food choices were being scrutinised. Meher, a first-generation Pakistani high-income homemaker married to a professional, expressed her frustration at Asian food being thought of as unhealthy:

> So I said that we don't eat parathas every day. . . . Like you eat Weetabix so we use wheat to make roti which is healthy. We use yogurt and salad. . . . Some people make food really spicy, so maybe that's why they think that what we eat it's not healthy.

Some others thought that healthy eating courses were not for Asian people because they did not include everyday South Asian food. Kajal, a Gujarati high-income homemaker married to a professional and living in a middle-class neighbourhood, shared her views on attending a healthy eating course at her son's school:

> I had done a course in [name of son]'s school for healthy eating. It was about salt content and sugar content in food, fats and saturated fats, unsaturated fats etc. we should buy I mean . . . after reading the label we should compare the salt levels and sugar levels and things like that . . . then healthy recipes like how we can include salad in our food or like if we have cous-cous then add some salad to couscous . . . it was not for Asian people.

Some mothers thought it was not cooking that was a problem for individuals from South Asian backgrounds but wider issues such as money, employment and a poor neighbourhood. Soha explained her reasons thus:

> These cookery courses are not really helpful. This is how your life is, you can't change life anyway. It is the way it is, you can't change it. It's very hard to change life, you can't, it's too hard. And cookery courses, I don't think so, nobody needs it really. I don't believe in cookery courses, honest. Only for the English community, not Asian community.

She goes on to say that Asian food is cheaper and healthier:

> Even SuperScrimpers they come out expensive sometimes, we [Asian] are cheaper than them. We know all this basically anyway from, what they're learning now, the Asian community know that ages ago, they just don't want to use it because they got lazier, that's it.

Like Soha, most of the mothers felt that Asian food is healthy if cooked properly with less oil. Amira explained how she fed her family healthily using traditional food: 'So yesterday I made a chicken biryani . . . so it's kind of healthy because I put *some chickpeas in, I put some green peas in and I add a bit of chicken in and it's rice'.* At the same time, there were some mothers across all three classes who felt that Asian food was less healthy.

Most mothers expressed doubts about any long-term benefits accrued from the healthy eating promotion messages they had received from various sources. Two mothers had attended a course that focused on understanding nutritional and calorific information on food items. They only found it helpful for a small number of items because most of the ethnic food they purchased did not have this information. Furthermore, they did not buy many ready-to-consume food items and could control how much salt or oil they put in their food. Many mothers from across all occupations, especially professional and skilled backgrounds, referred to the internet or books as their main source of information. Ameena, a skilled worker and first-generation Pakistani mother married to a professional, cited her sources of information on healthy eating:

> *I already have information about what is good, how much does it cost, maybe you can call it research. Nowadays everyone has access to the internet and you can find out about healthy eating, what is good for kids' growth and everything.*

Contrary to their lack of desire to engage with practitioners and health professionals regarding healthy eating, several mothers from across a range of occupations, especially professionals and skilled workers, had made substantial financial investments in purchasing utensils or machines that enabled them to eat healthily. Some bought a set of pans worth £2000 that can be used to make traditional Asian food without using any oil. Others reported investing in expensive fryers and grills. TV was an important source of information and some mothers also referred to Weight Watchers and Slimming World as sources of information about healthy eating.

Discussion and conclusions

In this chapter, I examined the diverse meanings and practices of healthy eating and experiences of healthy eating initiatives among two South Asian groups in the UK: Pakistani Muslims and Gujarati Hindus. I explored these from social class perspectives through the narratives of mothers who were employed in a range of occupations and had one or more dependent children. My research built on recent research that critically examines the relevance of class distinctions and taste in contemporary times and argues for a new perspective that recognises the dynamic and fluid nature of taste and the extensive cultural capital employed by people from low-income backgrounds to achieve their desired nutritional outcome (Beagan, Chapman, and Power, 2016; Smith Maguire, 2016; Flemmen, Hjellbrekke, and Jarness, 2018). The novel contribution of this chapter is to illustrate the intersection of ethnicity, gender and social class in healthy eating among these mothers. Class-based practices were dynamic and fluid and thus a nuanced understanding of healthy eating practices across diverse occupations is required.

Mothers' narratives showed considerable diversity in terms of the meaning and value of healthy eating. Professionals and some high-income homemakers provided an expansive account of what they meant by healthy eating, which encompassed not only food categories but quality of food, manners, cuisine and rules around eating. Their narratives suggested the symbolic value of healthy eating practices in their households such as ideals of equality, egalitarian gender roles, intimacy between couples and parental ideologies. In contrast, low-income homemakers and those in skilled work provided a brief description of healthy eating that was often limited to the consumption of fruit and vegetables and a reduction in salt and oil; although they emphasised cooking and eating fresh food and expressed their dislike for convenience food more often than professional mothers. Notably, being able to articulate an expansive discourse about eating healthily did not necessarily translate into healthy cooking and eating practices, and there appeared to be no significant difference in terms of how healthily respondents ate across occupations. Mothers from all four occupational backgrounds and both ethnicities reported eating healthily as well as not eating healthily, although their reasons varied. Overall, both in cases of transnational as well as same generation marriages, Gujarati mothers reported more consensus with regard to the meaning and practices of eating healthily in their household than Pakistani mothers. Gujarati mothers often reported cooking vegetarian meals for the whole family without much dispute. Pakistani mothers reported more disparities in meaning of healthy eating within the household and as a consequence more than one meal being cooked. There are four distinct areas that necessitate a reconsideration of Bourdieu's theory to make it relevant in context of race/ethnicity.

Firstly, how class intersects with migration especially in context of transnational marriages needs to be considered. Some research has shown that second- and third-generation migrants differ from first-generations in terms of eating fewer traditional meals (Gilbert and Khokhar, 2008). However, my findings suggest a more nuanced approach is needed towards the way food changes over generations, including recognition of cyclicity in food practices. For example, in transnational marriages, a second- or third-generation person may marry a first-generation migrant and their food practices revert to being more traditional, as was seen in several of the narratives. In some of these marriages, husbands and wives may occupy different class positions in terms of their education, employment and possession of cultural capital and in these 'mixed class' households, as several Pakistani mothers had reported, several meals were cooked at the same time catering to different tastes.

Secondly, in addition to cultural and economic capital, the chapter shows influence of ethnicity on food choice in two ways: (1) Asian food is sometimes portrayed as versatile and healthy, at other times spicy and oily was presented in comparison with the mainstream British food which was considered the norm, healthy option (although often more expensive) sometimes despite its obvious fat content, such as fish and chips. This internalisation of Asian food as being bad and western food being healthy and legitimate needs to be considered alongside class-based practices (Qureshi, 2019). (2) Mothers often desired for their children to develop a taste for ethnic food irrespective of its nutritional content, to enable them to become a respectable member of their religoethnic community, as seen in previous research (Salway, Chowbey, and Clakre, 2009).

Thirdly, a need for reconsideration of cultural capital in healthy cooking is required. Many mothers from both Gujarati and Pakistani backgrounds were able to cook and eat healthily because they knew how to purchase cheap and fresh food for example from a van man or food market as opposed to super store and buying groceries in bulk. They had extensive knowledge of various low budget and versatile recipes and ways to cook from scratch, and had developed relatively advanced cooking skills over the years. This necessitates a reconsideration of how cultural capital in healthy eating food practices is conceptualised and enacted (Beagan, Chapman, and Power, 2016; Smith Maguire, 2016).

Fourthly, the need to consider effects of wider gendered and racial discrimination on household food practices. Those employed in demanding professions often made reference to long working hours and a lack of energy and time for cooking. Although time is often an issue with those working long hours and has implications for food practices as shown with other populations (Blake et al., 2011), the additional pressure of being a woman and from a racial minority occupying higher positions were expressed as a concern. This is not surprising considering the gendered and ethnic labour market inequalities (Nazroo and Kapadia, 2013). Time appeared to be a major factor for some high earners who, compared to some manual workers and homemakers (both low income and high income) who were more often able to adopt strategies that allowed them to eat healthily on a budget, more often reported cooking and eating less healthily despite of possessing high levels of economic and cultural capital.

The findings challenge the assumptions enshrined in healthy eating initiatives, such as that South Asian populations are not already versed in healthy eating messages and that there is a lack of diversity in the meaning of food practices among them. These initiatives fail to recognise temporality and cyclicity in food practices, a lack of appreciation of the cultural capital displayed in advanced cooking skills and the ability to cook food on a budget, and assume that healthy eating messages and initiatives such as healthy eating courses will engender transformative change. The findings suggest a more nuanced understanding of individual and familial circumstances and socioeconomic location is required to engage with individuals and communities from diverse ethnic and social backgrounds. There are also wider factors related to affordability, availability of healthy food, access to relevant and accessible information about healthy eating and its impact on food practices within households. To engender transformative changes in food practices, these need to be addressed simultaneously alongside initiatives focused on enhancing the capabilities of individuals, families and communities to eat healthily.

References

Anderson, A.S., Bush, H., Lean, M., et al. (2005). Evolution of atherogenic diets in south Asian and Italian women after migration to a higher risk region. *Journal of Human Nutrition & Dietetics*, 18(1), pp. 33–43.

Archer, L. (2011). Constructing minority ethnic middle-class identity: An exploratory study with parents, pupils and young professionals. *Sociology*, 45(1), pp. 134–151.

Beagan, B.L., Chapman, G.E. and Power, E.M. (2016). Cultural and symbolic capital with and without economic constraint. *Food, Culture & Society*, 19(1), pp. 45–70,

Blake, C.E., Wethington, E., Farrell, T.J., et al. (2011). Behavioral contexts, food-choice coping strategies, and dietary quality of a multiethnic sample of employed parents. *Journal of the American Dietetic Association*, 111(3), pp. 401–407.

Bourdieu, P. (1984). *Distinction: A social critique of the judgement of taste*. London: Routledge and Kegan Paul.

Bourdieu, P. (1987). What makes a social class? On the theoretical and practical existence of groups. *Berkeley Journal of Sociology*, 32, pp. 1–17.

Bourdieu, P. (2011). The forms of capital. *Cultural Theory: An Anthology*, 1, pp. 81–93.

Bradby, H. (1997). Health, eating and heart attacks: Glaswegian Punjabi women's thinking about everyday food. In: P. Caplan, ed., *Food, health and identity*. London and New York: Routledge, pp. 213–233.

Charmaz, K. (2006). *Constructing grounded theory: A practical guide through qualitative analysis (Introducing qualitative methods series)*. London: Sage.

Chowbey, P. (2016). Employment, masculinities, and domestic violence in 'fragile 'contexts: Pakistani women in Pakistan and the UK. *Gender & Development*, 24(3), pp. 493–509.

Chowbey, P. (2017). What is food without love? The micro-politics of food practices among south Asians in Britain, India, and Pakistan. *Sociological Research Online*, 22(3), pp. 165–185.

Chowbey, P. and Harrop, D. (2016). *Healthy eating in UK minority ethnic households: Influences and way forward*. Discussion Paper. Race Equality Foundation.

Chowdhury, A.M., Helman, C. and Greenhalgh, P.M. (2000). Food beliefs and practices among British Bangladeshis with diabetes: Implications for health education. *Anthropology & Medicine*, 7(2), pp. 209–226.

Dale, A. and Ahmed, S. (2011). Marriage and employment patterns amongst UK-raised Indian, Pakistani, and Bangladeshi women. *Ethnic and Racial Studies*, 34(6), pp. 902–924.

Daye, S.J. (2016). *Middle-class blacks in Britain: A racial fraction of a class group or a class fraction of a racial group*. New York: Springer.

Devine, C.M., Connors, M.M., Sobal, J., et al. (2003). Sandwiching it in: Spillover of work onto food choices and family roles in low-and moderate-income urban households. *Social Science & Medicine*, 56(3), pp. 617–630.

Devine, C.M., Jastran, M., Jabs, J., et al. (2006). "A lot of sacrifices:" Work – family spillover and the food choice coping strategies of low-wage employed parents. *Social Science & Medicine*, 63(10), pp. 2591–2603.

Emadian, A., England, C.Y. and Thompson, J.L. Dietary intake and factors influencing eating behaviours in overweight and obese South Asian men living in the UK: mixed method study. *BMJ Open* 2017;7:e016919. doi: 10.1136/bmjopen-2017-016919.

Flemmen, M., Hjellbrekke, J. and Jarness, V. (2018). Class, culture and culinary tastes: Cultural distinctions and social class divisions in contemporary Norway. *Sociology*, 52(1), pp. 128–149.

Gilbert, P.A. and Khokhar, S. (2008). Changing dietary habits of ethnic groups in Europe and implications for health. *Nutrition Reviews*, 66(4), pp. 203–215.

Jamal, A. (1998). Food consumption among ethnic minorities: The case of British Pakistanis in Bradford, UK. *British Food Journal*, 100(5), pp. 221–233.

Kerrane, K.E. (2017). *Negotiating gender identity, motherhood and consumption: Examining the experiences of south Asian women in the UK*. PhD thesis. The Open University.

Lawrence, J.M., Devlin, E., MacAskill, S., et al. (2007). Factors that affect the food choices made by girls and young women, from minority ethnic groups, living in the UK. *Journal of Human Nutrition and Dietetics*, 20(4), pp. 311–319.

Lawton, J., Ahmad, N., Hanna, L., et al. (2008). 'We should change ourselves, but we can't': Accounts of food and eating practices amongst British Pakistanis and Indians with type 2 diabetes. *Ethnicity & Health*, 13(4), pp. 305–319.

Lofink, H.E. (2012). 'The worst of the Bangladeshi and the worst of the British': Exploring eating patterns and practices among British Bangladeshi adolescents in East London. *Ethnicity & Health*, 17(4), pp. 385–401. doi:10.1080/13557858.2011.645154.

Ludwig, A.F., Cox, P. and Ellahi, B. (2011). Social and cultural construction of obesity among Pakistani Muslim women in North West England. *Public Health Nutrition*, 14(10), pp. 1842–1850.

Maylor, U. and Williams, K. (2011). Challenges in theorising 'Black middle-class' women: Education, experience and authenticity. *Gender and Education*, 23(3), pp. 345–356.

Modood, T. (2004). Capitals, ethnic identity and educational qualifications. *Cultural Trends*, 13(2), pp. 87–105.

Nazroo, J.Y. and Kapadia, D. (2013). Ethnic inequalities in labour market participation? Available at: www.ethnicity.ac.uk/medialibrary/briefingsupdated/Ethnic%20inequalities%20in%20labour%20market%20participation.pdf [Accessed 28 July 2016].

Pallan, M., Parry, J. and Adab, P. (2012). Contextual influences on the development of obesity in children: A case study of UK South Asian communities. *Preventive Medicine*, 54(3–4), pp. 205–211.

Phoenix, A. (2006). Interrogating intersectionality: Productive ways of theorising multiple positioning. *Kvinder, Køn & Forskning*, pp. 2–3.

Public Health England and the Food Standards Agency (2014). *National diet and nutrition survey: Results from years 1,2,3 and 4 (combined) of the rolling programme 2008/2009–2011/2012*. Available at: www.gov.uk/government/statistics/national-diet-and-nutrition-survey-results-from-years-1-to-4-combined-of-the-rolling-programme-for-2008-and-2009-to-2011-and-2012

Qureshi, K. (2019). *Chronic illness in a Pakistani labour diaspora*. Durham, NC: Carolina Academic Press.

Rollock, N., Vincent, C., Gillborn, D., et al. (2013). 'Middle class by profession': Class status and identification amongst the Black middle classes. *Ethnicities*, 13(3), pp. 253–275.

Roos, E., Sarlio-Lähteenkorva, S., Lallukka, T., et al. (2007). Associations of work – family conflicts with food habits and physical activity. *Public Health Nutrition*, 10(3), pp. 222–229.

Salway, S., Chowbey, P. and Clakre, L. (2009). Parenting in modern Britain: Understanding the experiences of Asian fathers. Available at: www.jrf.org.uk/sites/files/jrf/Asian-fathers-Britain-full.pdf

Sevak, L., Mangtani, P., McCormack, V., et al. (2004). Validation of a food frequency questionnaire to assess macro-and micro-nutrient intake among south Asians in the United Kingdom. *European Journal of Nutrition*, 43(3), pp. 160–168.

Smith, M.J. (2016). Introduction: Looking at food practices and taste across the class divide. *Food, Culture & Society*, 19(1), pp. 11–18. doi: https://doi.org/10.1080/15528014.2016.1144995

Song, M. (2003). *Choosing ethnic identity*. Cambridge, UK and Malden, MA: Polity Press.

Williams, R., Bush, H., Lean, M., et al. (1998). Food choice and culture in a cosmopolitan city: South Asians, Italians and other Glaswegians. *The Nation's Diet, The Social Science of Food Choice*, pp. 267–286.

Williams, S. (1995). Theorising class, health and lifestyles: Can Bourdieu help us? *Sociology of Health and Illness*, 17, pp. 577–604.

Wyke, S. and Landman, J. (1997). Healthy eating? Diet and cuisine amongst Scottish South Asian people. *British Food Journal*, 99(1), pp. 27–34.

Chapter 8

The original taste of real food

The discursive formation of Taiwan's food education

Ming-Tse Hung

Introduction

The chapter starts with a short introduction to the food safety scandals in Taiwan and explains how the question of real food was raised afterwards, leading to the development of food education in Taiwan. As the Council of Agriculture (CoA) claims, 'it is only through the promotion of food education can we make more people realize the importance of real food, bring the real taste of food back to our dining tables, reacquaint the land we live, and reestablish our relationships with agricultural workers' (CoA, 2017, p. 11).

Adopting the Foucauldian approach of discourse analysis, this chapter examines 57 teaching plans from the Agriculture and Food Agency, proposals for Agro-Food Education Basic Law and relevant news reports to understand how the question of real food has been raised and answered. Next, the problematisation of taste is examined, specifically the notions of loss of food's 'original taste' and eaters' inability to properly taste food. As the fakeness of food is interpreted as a sensual object that can be identified by the eater's senses of smell and taste, it entails the training of taste and the requirement for more direct experience with food. Lastly, this chapter examines the emergence of food producers and farmers as the speaking subject of food education discourse, explaining that the pursuit of originality and realness of food authorises them to advise on the identification of real food.

From food safety to food education

Since 2008 there have been three major food safety scandals in Taiwan: the 2011 abuse of plasticisers, the 2013 contaminated starch scandal and the 2014 gutter oil scandal (Taiwan Food Safety Summit, 2016). In 2011, a laboratory staff member from the Taiwan Food and Drug Administration (TFDA) by chance found DEHP (Di (2-ethylhexyl) phthalate) in a probiotic powder; after investigations, more foods were found to contain DEHP as well as other plasticisers. The TFDA discovered that some food-additive distributers had illegally added the banned industrial plasticiser to a stabilising emulsifier as a cheap substitute for palm oil.

The resulting product, a clouding agent, was then used by many famous food manufacturers in juices, sports drinks and other beverages. A total of 426 food producers were involved in the scandal and 965 food products were pulled from nearly 40,000 shops in Taiwan (Taiwan Food Safety Summit, 2016).

In 2013, health authorities announced that they had found some foods in local markets contained maleic anhydride, an industrial material used mainly to produce polyester resin and pesticides, and not authorised for use in foods. The investigation eventually tracked down six producers making the tainted modified starch and found that about 19.11 million kilogrammes was sold to food factories and restaurants. In 2014, 645 tons of adulterated cooking oil, often referred to as 'gutter oil', were found to have been produced and distributed to more than 1200 restaurants, schools and food processors in Taiwan, more than 1300 food products contaminated by the oil and hundreds of tons of products were removed from shelves. The company Chang Guann purchased waste oil collected from restaurant fryers, sewer drains, grease traps and slaughterhouse waste, filtered, boiled and refined it, and mixed it with lard to make its Chuan Tung cooking oil as a cheaper alternative to normal cooking oil (Taiwan Food Safety Summit, 2016). The Taiwan Food Safety Summit (2016) claims that it is getting more and more difficult for the public to know what they consume on a daily basis and suggests that the authorities set the standards and develop appropriate testing methods, otherwise 'it is not possible to find out which food is fake and which is real'.

In a feature report on Taiwan's food safety issues, the editor of the magazine *Global Views Monthly* raised the question of edibility: 'From DEHP, the leanness-enhancing agents in pork to the tainted modified starch, there have been several food safety scandals in recent years, and now even the cooking oil is adulterated, which makes us wonder what on earth is edible?' (Wang, 2013). A report about the gutter oil scandal also expresses a similar idea, arguing that almost everyone in Taiwan having eaten something made with inedible gutter oil (Peng, 2014). The scandals raised a general doubt about edibility. No matter what they are, or when, and where they are consumed, foods are potentially toxic and inedible. Edibility, the defining characteristic of food, is put into question; 'real' in the context of these scandals is tightly connected to the property of being fit to eat. 'Real food' is to a certain degree a tautology, used to reconfirm edibility and reinforce the definition of food as something fit to eat. The notion of 'inedible food' is made possible in the event of food safety scandals. Instead of saying that contaminated food is 'not food', reports on the scandals say it is 'fake'. The replacement of food/non-food with real/fake food turns the external contrast into an internal one: food can be either real or fake, edible or inedible.

Fake food is an ambiguous concept; it lies in the grey zone between edible and inedible, the adjective pointing out the inedibility while the noun implies something still fit to be eaten. The ambiguity also comes from the simple fact that these products contain both edible and inedible ingredients. As the inedibility is not able to be identified directly by observation, and the existing food tests failed to warn against the illegal additives in advance, food education acquires an important role here as a more fundamental solution, and the real/fake food distinction is passed from food safety scandals to food education, forming a discursive field to produce and circulate its own knowledge about food.

Following the wave of scandals, the government vowed to take proactive action, strengthen the inspection of staple food products and lay out more severe fines for wilful violations of the law (Chung, 2013). The government increased its food safety budget and worked on amendments to the Act Governing Food Safety and Sanitation to strengthen existing regulations and ensure food producers comply with them. But some experts warned that Taiwan's food problem cannot be solved unless people start being given the right knowledge about food. Voices called for food education measures to be implemented to enhance people's understanding of what they eat and re-establish their relationship with food.

Civil groups play an important role in promoting food education. In 2011, the Homemakers United Foundation (HUF) started its 'Green Food Education' project, the first food education programme in Taiwan, to address topics including genetically modified food, the reduction of pesticide use and food literacy at home and school (HUF, n.d.). In 2014, the Agro-food Education Legislation Promotion League was established and proposed a draft law to the Legislative Yuan suggesting cooperation between governments, schools and communities to 'bring real food back to Taiwan' (HUF, 2015). In 2016, the Taiwan Association for Food Education was formed and joined by reputable entrepreneurs, chefs, writers and journalists aiming to promote local and seasonal produce and provide food education for children (Teng, 2017).

Local governments in Taiwan have also been engaging in developing their own food education projects. From 2012 to 2014, half of Taiwan's counties and cities started their food education projects (Yung, 2018). Since 2013, the Taipei City Department of Economic Development (DEDT) has been maintaining a platform for food education, training seeded teachers, editing the *Taipei Agro-food Education Handbook* and supporting cooperation between organic farms and local primary schools to develop food education classes (Yung, 2018). Kaohsiung Education Bureau has also published 3 volumes of food education textbooks which introduce 12 local foods in detail (Lu, 2017). The Tainan City Government announced that 'food education' would be added to the five traditional elements of a well-rounded education – moral, intellectual, physical, social and aesthetic – and become part of the curriculum (Yung, 2018); it published 5 textbooks themed around Tainan's local foods in 2017.

Following the example of Japan's Basic Law on Shokuiku (Food Education) adopted in 2005, the Legislative Yuan has been pushing for a Food Education Basic Law in Taiwan (Yung, 2018). In 2012, the Legislative Yuan held a public hearing to discuss the draft law, aiming to incorporate resources and to provide an overall structure. Though the bill is still in the drafting stage, two local governments have legislated their food education acts. In 2015, the Ilan government passed the 'Ilan Self-Government Ordinances of Healthy Diet', requiring every primary and secondary school in Ilan to establish a school farm and include food education lessons in the curriculum (Yung, 2018). In 2018, the Taichung City Government passed the 'Taichung Self-Government Ordinances of Agro-food Education', aiming to integrate agriculture into the curriculum and promote food safety education via experiential activities and participation (Lu, 2018).

Alongside the engagement of local governments and civil groups, the central government authorities contribute to the promotion of food education by developing teaching materials and plans for schools' reference. In 2012, the Council of Agriculture conducted the first research on the design of agro-food education, and one year later the Agriculture and Food Agency (AFA) launched the Organic Agro-food Education programme, conveying an organic agro-food education to junior high and primary-school teachers at conferences. To encourage teachers to promote such education, AFA held its first teaching plan contest and selected 57 entries to share with teachers (AFA, 2017). Since then, several contests have been held and more teaching plans shared by the AFA and other authorities.

These initiatives and projects represent a distinct way of knowing food from nutrition science, the paradigm that looks at food as being composed of nutrients and encourages eaters to understand food at the biochemical level (Scrinis, 2008). But what distinguishes food education from nutrition science? To understand the differences, this research examines the 57 teaching plans collected by the AFA, the proposals of Taiwan's Agro-Food Education Basic Law, and print and online articles published around food education since 2010 – this is when food education started to develop and to attract public attention. The analysis focuses on two aspects: the discursive object of food education and the speaking subject that is allowed to have its voice. As Foucault indicates,

> Knowledge is . . . specified by that fact: the domain constituted by the different objects that will or will not acquire a scientific status . . . knowledge is also the space in which the subject may take up a position and speak of the objects with which he deals in his discourse.
>
> (Foucault, 2004, p. 201)

First, what object can be identified in food education discourse? What is there to be known and discussed about food? How do we know it? Unlike nutrition science, food education does not rely on scientific experiments, assessment and measurement to produce the nutri-biochemical level of knowledge, but rather emphasises on the direct experience with food to define 'real' and 'fake', a question that cannot be answered by objectifying food as composition of nutrients. If not the biochemical structure and nutritional functionality of food, what does food education discourse see as its object? As knowledge is inseparable from the procedure establishing it (Deleuze, 2006), the above inquiry leads to the question of practice: to access the object that allows the distinction between real and fake, what practice, process and method are deployed?

Second, since food education as an education system also involves 'a qualification and a fixing of the roles for speaking subjects' (Foucault, 1981, p. 64), we must ask,

> [W]ho is speaking? Who . . . is accorded the right to use this sort of language? Who is qualified to do so? . . . What is the status of the individuals who – alone – have the right, sanctioned by law or tradition, juridically defined or spontaneously accepted, to proffer such a discourse?
>
> (Foucault, 2004, p. 55)

The status of the speaking subject involves 'criteria of competence and knowledge; institutions, systems, pedagogic norms; legal conditions that give the right' (Foucault, 2004, p. 55). For nutrition science, one must pass the dietitian examination to hold a valid license and receive continuing education to renew it; it is the certification and legal regulation that justify the dietitian's position to speak. As there has been no such formal and institutional system to form the speaking position of food education, it is necessary to ask: Who is qualified to speak in the discourse of food education? On what conditions? It should be noted that the inquiry into the speaking subject is not about looking for a transcendental subject, collective consciousness or a psychological subjectivity that defines the domain. Rather, it is about describing 'an anonymous field whose configuration defines the possible position of speaking subjects' (Foucault, 2004, p. 137), examining the context and identifying the criteria that need to be met for one to speak. In short, the research questions are as follows: What discursive object is identified in food education discourse? With what practices can one know the object? Who is allowed to tell the truth about food? On what conditions?

Problematisation of taste

After the food safety scandals, the uncertainty of edibility became a common concern in Taiwan. This uncertainty was connected not only to the illegal additives but to additives in general, generating discussion around real/fake food, foods with additives being deemed fake and inauthentic, and additive-free foods real and pure. As Ulloa (2018) rightly points out, 'The artificial has strongly become synonymous with the fictitious and the simulated, and consequently the antonym of the genuine and real' (Ulloa, 2018, p. 66). This dichotomy might be considered a way of addressing the issue of uncertainty, or at least as an attempt to locate it after the failure of the existing food safety system and the additional regulations aiming to tackle food safety problems.

In discussions around the real/fake food distinction, taste is often identified as one of the key indicators. For instance, the Homemakers United Foundation (HUF), the first NGO promoting food education in Taiwan, provides a very clear explanation: 'real' means that the food product is produced with the actual food material; 'fake' means the taste and flavour are synthesised with chemicals and additives (Tu and Huang, 2016). For industrial foods, a desired flavour profile is achieved for a specific food or beverage product by the blending of approved chemicals and extracts; the tastes are thus deemed alienated and no longer emerge from the food itself. As flavourists imitate natural flavours by using chemicals whose odour and flavour are similar, unlimited tastes can be achieved without the need to use real ingredients. The missing link between the taste and the ingredient is what defines fake foods. On the contrary, real foods are those comprising 'real' ingredients, whose tastes/flavours can be properly associated: 'We should keep the original taste a vegetable has in its nature, no more and no less to call it real food' (Wu, n.d.). 'Real' in this sense means 'adequate', asserting that the appearance, taste or physical condition of the food emerges from nature without artificial intervention.

Behind the attempt to distinguish real and fake foods is the blurring of boundaries between them. There is no need to stress the real/fake distinction if the differences are obvious enough to spot. Only when the fake is indistinguishable from the real is a means of telling them apart needed. In a news conference held by legislator Wang, Yu-Min and the NGO John Tung Foundation, they expressed their concerns that many children think the zero-fruit juice drinks produced with flavourings and additives taste fresher than the natural ones (Sun, 2012). One news article makes the same observation, saying that 'with a drop of fruit flavouring . . . the ice cream and jam look real and genuine with absolutely no fruits in them, but the consumers' tongues cannot tell' (Chen, 2014). It concludes that consumers are not merely incapable of telling the taste of natural juices from that of fruitless drinks but consider the real juices as spoiled and the artificially flavoured ones fresh (Chen, 2014). For consumers, the flavours created by flavourists taste just like the original foods and become '"second nature" to us' (Ulloa, 2018, p. 70).

It should be noted that confusion about tastes is not exclusive to the discourse around real/fake food and existed before the food safety scandals, only with a different focus. On 28 May 2010, the Taiwan Food and Drug Administration (TFDA) published a regulation on the labelling of instant noodles, and on 20 July 2015 on the labelling of freshly made beverages in chain convenience stores and fast food stores (TFDA, 2010, 2014). The Food Safety and Sanitation Act Chapter V, 'Food Labelling and Advertisement', Article 22, requires that food product names must be unmistakable. All packaged instant noodles that include only seasonings or flavourings are required to bear on their outer packaging the term 'Flavoured Noodles'. Two years later, the same rule was extended to apply to tea and juice drinks. The Food and Drug Administration announced that after July 2014, all packaged fruit and vegetable juices would be required to indicate on the outer package the percentage of natural juice. Only those made of 100 percent natural juice might be labelled 'pure juice'; those drinks flavoured only with flavourings would be required to be labelled 'xx flavour' or 'xx taste' (Shi, Z.Y., 2014).

The above regulations problematise the correspondence between names and ingredients and assert that the proper solution is clearer labelling, which is only necessary when eaters might otherwise mistake the tastes of flavourings for the tastes of actual ingredients. Clarifying the distinction between synthesised and real ingredients unmasks the absence of real food ingredients, but food education addresses this question in another way. Unlike mandatory labelling requirements, food education sees the confusion as a confusion of taste; the failure of the distinction is not about naming but about tasting – it takes not only eyes but also tongues to really know food. The problematisation of taste includes two aspects: the loss of food's 'original' taste, and the eater's inability to properly taste food. To a certain degree, the two reinforce each other: the exposure to additives numbs eaters' tongues, while this numbness encourages the growing use of additives. In a food forum held by Taipei City Office of Commerce that appealed to the public to critically engage with food, the chairman of Tanhou Food said:

> People eat out all the time and consume those foods with lots of additives; the original tastes of real foods can no longer be tasted. . . . [M]ost people's tongues have been numbed by colourants, flavourings, and sweeteners, and real foods with no additives are thought to be tasteless.
>
> (Huang, 2016)

It is argued that eaters often consider real foods as plain and tasteless, as they do not have the ability to appreciate the original tastes (see e.g. Zhu, 2013; Qiu, 2015; Zhang, 2016), and their tongues are numbed by the additives so widely used in food products. The author of *From Appetite to Food Education* argues that this ability was lost long ago in modern times, as fast foods and processed foods came to dominate the market, most people eating food products in which the ingredients are not recognizable (Qiu, 2015). Qiu calls this development 'food illiteracy' and thinks it is inevitable in the era of industrial food, where people are alienated from the land and do not know where food comes from, and where highly processed foods with additives destroy children's sense of taste and appetite for real foods (Qiu, 2015). The same idea is expressed in different places. Leezen, one of the main organic chain stores in Taiwan, claims that if consumers consume food products with chemical ingredients over long periods, their sense of taste may degenerate; they may memorise the tastes of these chemicals but not those of the foods and no longer be able to appreciate natural foods. For this reason, Leezen claims to promote a 'real food diet', featuring natural and real foods with the original tastes (Leezen, 2016). *EverydayHealth*, one of the most widely read health magazines in Taiwan, expresses the same idea, stating that 90 percent of the Taipei population has already lost the ability to taste food; unhealthy diets are gradually numbing people's tongues. The article claims that this is an age of taste degeneration: more and more flavourings are added to make food products appetizing, but as a result, people are losing the ability to appreciate real tastes (EverydayHealth, 2014).

Impaired taste is perceived as a critical issue to be addressed in discourse around real food, but it is problematised in a different way from medical discourses, which consider impaired taste as primarily a symptom of other diseases; causes of impaired taste range from the common cold to more serious medical conditions involving the central nervous system (Shi, J.R., 2014). To restore the patient's taste, the underlying condition should first be treated. Bacterial sinusitis, salivary glands and throat infections can be treated with antibiotics, and symptoms of colds, flu and allergic rhinitis that impact taste may be relieved with decongestants or antihistamines (Shi, J.R., 2014). Here, treatment is targeted at the sickness that causes impaired taste – not the sense of taste itself. In the discourse of real food, however, impaired taste in itself is the issue to be addressed. An improper correspondence between taste and food and an incapacity to notice the absence of real ingredients are deemed to be the cause of various sicknesses. For instance, Leezen claims:

> The consumers' senses of taste are deceived, and diseases of civilization come to us one by one. . . . To get the real original taste of food back and keep our tongues from being deceived by fake foods full of additives, we must understand the nature of food.
>
> (Chen, 2017, p. 5)

The notion of impaired taste as a cause of diseases also finds a place in medical discourse and therapies have been developed to train patients' sense of taste to help improve their health conditions. In 2011, Jen-Ai Hospital proposed the idea of Taste Quotient (TQ) and opened the TQ Training Centre in Taichung, Taiwan. TQ is defined as 'the ability and intelligence to taste, or the ability and intelligence that can be trained or improved by tasting' (Liu, 2011). The TQ Training Centre claims that evaluating, testing and training the sense of taste successfully helps obese patients lose weight and improves other diet-related sicknesses; it provides courses to train patients' ability to identify tastes and to memorise, communicate, learn, balance and control them, to help patients develop better diet habits. Though the notion of TQ does not link to food education directly, both concepts problematise the sense of taste and, in the broader discursive context, facilitate the development of a sensual experience-centred knowledge of food.

Training of taste

> Most people don't know the real taste of food. . . . [I]f they can't identify the source of the real taste, it may result in the overconsumption of fake foods. . . . [T]he only way to understand the real taste is to 'widen the experience of taste'.
>
> (Wang, 2015)

To address alienated tastes and impaired senses, food education promoters propose to enhance the sensual experience and encourage eaters to distinguish the real from the fake. For instance, Mini Cook, the first children's cookery studio in Taiwan, writes that 'through preparation and cooking, children get to touch real foods and experience the real tastes. . . . [F]ood education is about teaching not a skill but an ability to interact with food' (Mini Cook, n.d.). In one of the games the studio designs to help children better know their foods, children are asked to wear eye patches and identify fruits by their smells. Mini Cook thinks the game helps kids 'memorise the real flavours of foods . . . and distinguish natural foods from artificial flavourings' (Mini Cook, 2015, p. 90). One of New Taipei City's 2015 food education teaching plans also mentions that food education 'should activate students' ability to observe, smell, hear, touch . . . [and to] experience the original look and taste of food' (New Taipei City Government, 2015, p. 17). Similarly, the Agro-Food Education Basic Law defines food education as 'a learning-by-doing process for students to find out about local agriculture and proper diet' (Draft of the Agro-Food Education Basic Law, 2015, p. 48).

The training of taste to distinguish the real from the fake constitutes a major part of the teaching plans collected for this research. The 'Expose the Additives' teaching plan, for instance, aims to 'teach children how to identify real food and choose good food, to lower their consumption of processed food products and eat more natural food' (Chang, 2014, p. 2). For this purpose, it designs an activity for the students to taste and compare several sets of food items, each including two samples – one with additives and one additive-free. Through the practice of tasting, the taste of food is enabled to enter into the game of truth and becomes an object for thought (Foucault, 1988), as well as an object for verification in food education. First of all, it is necessary to identify a lack of 'real' ingredients in foods in order to raise the relationship between tastes and ingredients as a question to address.

> Ask students to collect packaging from cookies, candies, and breads and analyse the lists of ingredients on them: 1. Look carefully at the ingredients of most of the cookies on the market – there are no natural ingredients listed for the flavours they claim to have. These flavours mostly come from the artificial flavouring agents.
>
> (Lin, 2013, p. 4)

As discussed earlier, this approach indicates only the disconnection between names and ingredients; it is the taste of food that is the ultimate object for the distinction between real and fake foods. In several teaching plans, students are asked to identify the taste of artificial flavourings and compare it to that of 'real foods', so the absence of real ingredients can be unmasked. The 'Food Additives' teaching plan clearly states that fruit-flavoured candies are 'fake' (Food Education Teaching Plan Editing Team, 2013, p. 41) because of the lack of actual fruits. After practising tasting and comparing real and fake foods, students are supposed to be able to tell the difference between natural and artificial flavours and tastes.

1. Smell the flavourings. Compare the tastes of fruit-flavoured candies and real fruits. . .

2. Take steps such as smelling grape flavouring, tasting grape-flavoured candies, eating a grape
3. Students compare the three items
                    (Food Education Teaching Plan Editing Team, 2013, p. 34)

Further, the invisible existence of the flavourings could be made visible through the demonstration of the juice drink blending. The 'Find Out Where the Colours Come From' teaching plan presents the additives for observing and tasting to 'let students understand the original tastes of foods, and avoid the risks of chemical additives':

Discover Orange Soda

1. Open the bottle of orange soda. Let students smell the flavour and guess if there's orange juice in it
2. Peel oranges and let students smell, observe, and compare oranges and soda drinks
3. Ask students to guess again if there's orange juice in the soda drinks
4. Present the chemical additives in artificial juice drinks. Let students smell the flavourings, make one glass of orange soda, and ask students to smell and observe it again.

(Lin, 2013, p. 2)

Via showing how flavourings are added to make juice drinks, the link between flavourings and the ingredients they attempt to imitate is broken, revealing the absence of 'real' foods. By illustrating the existence of additives, the teaching plan aims to teach students how to avoid the risks of fake food, as fake food is deemed not fit to be eaten. The 'Food Safety Examined, Artificial Additives Out' teaching plan (Lin, 2013) explains in detail why artificial additives are inedible:

Activity

. . .
3. Artificial flavourings and colourings are extracted and refined from petrochemicals. As we know, the smell at the gas station is not pleasant and can even make people dizzy. Since gasoline is inedible and toxic, are the products extracted from it edible?
. . .
5. Buy organic, unflavoured steamed buns and share them with the students; let them taste the differences between additive-free foods and those with artificial additives.

(Lin, 2013, p. 4)

Evaluation

When I mentioned that the artificial additives are extracted from petro-chemicals . . . I added that 'eating these chemical additives is just like putting the fuel truck nozzle in your mouth and drinking gasoline'. . . . Though the steamed buns are not delicacies, they are pure and additive-free.

(Lin, 2013, p. 7)

The teaching plan starts by indicating the lack of natural ingredients in the industrial foods and later moves to the persistent inedibility of fake and artificial ingredients. The mechanism of food processing is simplified to an equation: if the material is inedible, the extraction or the product is equally inedible. From food safety scandals to food education, the inedibility of additives is continually generalised; it is not simply about whether the additives are used legally or illegally but about the very nature of the production process and the fact that these substances are not collected, grown or harvested in nature. The inedibility of artificial additives is also illustrated in 'The Witch's Secret Recipe' teaching plan.

Picture storybook: The Witch's Secret Recipe. The witch Mila tries to save the trouble of making juice in the traditional way and blends flavourings and colourings to make artificial juice, but the villagers have too much of the artificial drink and all get sick.

Activity

1. Teacher asks: Why does Mila's juice make the villagers sick? (If the children are unable to answer, repeat the part of the plot where Mila makes "fake" juice with flavourings, colourings, and sugar)
2. Ask children: Are flavourings and colourings edible?
3. Use a cotton swab to pick up the flavouring and let children smell it (unlike natural flavourings, the artificial flavourings smell pungent, as they are highly concentrated). Ask children: Do the flavourings smell edible?

(Children Agro-food Education Team, 2013, p. 2)

The pungent smell indicates the existence of artificial additives and inedibility, which is implied to be the cause of the villagers' sickness. Unlike in nutritional or medical discourse, the sickness is attributed to the consumption of the fake juice drink without any further explanation of the mechanism, and the intangible inedibility is directly exposed to primary experience. Another contrast established in this teaching plan is traditional/chemical, which is often synonymous with real/fake. Another teaching plan 'Distinguish Between the Real and Fake Soy Sauces' reveals this contrast, claiming to train students' ability to tell the differences between traditional soy sauces and artificial/chemical ones (Huang, 2014). Students are provided with 'real and fake soy sauces' (Huang, 2014,

p. 2) and asked to identify the differences in main ingredients, colours, taste and appearance when shaken (Huang, 2014, p. 6). 'Real' means remaining the original taste without any concealment or alteration, and following the norms, traditions and natural order, so taking shortcuts by using additives to simulate the tastes or shortening the production procedure with chemical agents is deemed 'fake' and should be rejected.

Farmers as the speaking subject

The originality of food as the object to know requires a speaking subject that is close enough to food, as it is deemed vulnerable to all kinds of interruptions and difficult to preserve. This closeness to food poses the first limitation for the speaking position, denying dietitians and doctors their once exclusive authority to teach the public what and how to eat. Instead, farmers are now given the right to advise on healthy diets: 'There is too much to learn about food, and one can never reach the end of it. Those who actually grow food must know much more' (Chen, 2016, p. 61). It is the formation of food education that grants farmers the speaking position and confirms their understanding of food. The proposal of Food Education Basic Law stresses the critical role farmers play, saying that 'learners should learn the local agriculture and proper diets through the interaction with . . . farmers, food workers and other relevant actors' (Draft of the Agro-Food Education Basic Law, 2015). The proper diets, which used to be considered the responsibility of dietitians or doctors to advise on, are now also something that can be learned from farmers and food workers.

It does not mean that the certificated professionals lose their voices altogether in the discourse of food education, but farmers do gain a significant raise on their volume. This recent change of farmers' status poses a sharp contrast. There was an old Taiwanese saying that those who can't study farm, which implies that farming does not require much knowledge or expertise. When the National Chi Nan University (2017) held a series of agri-food education activities, claiming that 'farms are classrooms, and farmers are teachers', it is not just a figure of speech but also indicates the fact that farmers, who were deemed to have nothing valuable to say and share, are now seen as the ones that can tell the public the truth about food.

Other than the closeness to real food, farmers' mode of life as a manifestation of the object's realness also entails their speech and justifies the speaking position. In Johnston and Baumann's (2009) analysis of authentic food, they argue that authenticity of the object is assured by the sincerity of the subject, an identifiable individual or group.

> To speak of the 'real' versions of things is to invoke the concept of authenticity. 'Simple' food is authentic because of . . . the association between authenticity and individual sincerity, or being 'true to oneself'.
>
> (Johnston and Baumann, 2009, p. 76)

My study of real food draws a similar conclusion that the realness of food is associated with the producer's mode of life. This association becomes an essential condition of truth-telling, a manifestation of truth and a standard for the truth-value of a statement. Whether certain food is real is not about the scientific evidence presented or the logical argument made, but about the producer's personal life, the background and the behaviours. Therefore, it is important to let the farmer speak, so we get to judge whether one is living a real life, producing real food and telling the truth.

> When searching for food ingredients, we care more about people than certification. Real food can only be found where there are real people . . . [who] make real food, and learn how to be true to others and oneself.
>
> (Bao Zi-Yi, 2018, pp. 86, 88)

This association between the object's realness and the subject's sincerity distinguishes the discourse of real food from nutrition science and medicine, where the truth-value of a statement is independent of the speaking subject and relies more on an institutionalised system. A dietitian or doctor first speaks as a member of the professional group, as someone who possesses the knowledge of nutrition science and medicine. It is not the background, experience, personality or lifestyle but the license or qualification that validates the speech, confirms the truth-value of the statement and grants the access to the truth. On the contrary, a farmer must express his own opinion, thought and conviction in his own name. To some degree, the lack of professional certification might make a farmer's speech more real and truthful, as with less institutional influence, one is deemed more likely to speak with one's personal voice from the life experience, which is seen as a more direct, intimate and pure experience without being distorted or affected.

Conclusion

Since the food safety scandals, the realness of food has been problematised by focusing on the inedibility of chemical or industrial ingredients, and the invisible yet potentially ubiquitous presence of inedible additives entails the training of taste and food education. Flavouring chemistry is deemed to alienate flavours from foods, and the construction of the lost 'original taste' of food as a discursive object facilitates the production and circulation of the knowledge of real food. From unknown to known, imperceptible to perceptible, the practice of tasting might be seen as a practice to manage the proliferated invisibility, uncertainty and riskiness of food. As Campbell (2013) states: 'Food constitutes one of the first kinds of broadly traded items implicated by a new politics of risk. . . . [F]ood was at the vanguard of products subject to the new management of that risk' (Campbell, 2013, p. 189).

Beck (1992) also indicates that the contamination of food, among the other forms of risk, generally remains invisible and initially only exist in terms of the knowledge about them (Beck, 1992, p. 23). Cooking and eating are becoming a kind of implicit food chemistry aiming to minimise the negative effects (Beck, 1992, p. 35), and without the scientific theories, experiments and instruments, such invisible and uninterpretable risks are beyond personal perception and experience.

> One no longer ascends merely from personal experience to general judg-
> ments, but rather general knowledge devoid of personal experience becomes
> the central determinant of personal experience. . . . [W]e are dealing not
> with 'second-hand experience', in risk consciousness, but with 'second-hand
> non-experience'. Furthermore, ultimately no one can know about risks, so
> long as to know means to have consciously experienced.
>
> (Beck, 1992, p. 72)

The intangibility of risks that only comes to consciousness in scientised thought cannot be directly related to primary experience (Beck, 1992, p. 52). However, by introducing the notion of 'original taste' and the taste of additives to put what escapes perceptibility back into the sensorial world, food's 'fakeness' becomes visible or interpretable. Inedibility, once only perceivable with theories, experiments and instruments, comes to be interpreted as a sensually experiential object, and the tasting of foods is used to provide a base of knowledge capturing what once escaped personal experience. Notions of the original taste of food and food education should be considered as an attempt to place general knowledge back into the realm of personal experience, to change 'second-hand non-experience' into 'first-hand experience', where food is known by being consciously experienced. In short, the move is to translate the unknown into the known.

The food education discourse claims that with proper training and practice one can become able to taste the invisible fakeness of food and avoid the risk of consuming inedible chemicals. This is a system of knowledge that is developed in parallel to nutritional science or medicine, which focuses on the biochemical level of food and entail the involvement of scientific help. The nutrition-based episteme treats the eater as unknowledgeable about themselves and places them in the position of being told how and what to eat by an expert (Scrinis, 2008). In contrast with nutrition science, food education discourse considers the eaters not so much 'unknowledgeable' as 'inexperienced'. As argued earlier, food education establishes itself on embodied experiences and sees food primarily as a sensual object but not something to be scientifically analysed. Correspondingly, a farmer as the major speaking subject of food education is less of a knowledgeable professional who has the exclusive access to truth than of an experienced individual knowing how to live a real life in terms of the relationship with food. With the association between the object and subject, the farmer's speech includes at least two aspects: annotation of the originality of food and the confession of the farmer's life, both of which are used to confirm the realness of food without relying on a scientific system of measurement.

References

Agriculture and Food Agency (AFA) (2017). *Note of the teaching plans*. Available at: www.afa.gov.tw/cht/index.php?code=list&ids=603 [Accessed 8 Feb. 2018].

Atkins, P. (2012). Social history of the science of food analysis and the control of adulteration. In: A. Murcott, W. Belasco, and P. Jackson, eds., 2013. *The handbook of food research*. London: Bloomsbury, pp. 97–108.

Bao, Z.Y. (2018). New Taipei – process-intensive grain drinks. *Gradually Move on with Honest Food* III, 27 Mar., pp. 86–89.

Beck, U. (1992). *Risk society: Towards a new modernity*. London: Sage Publications.

Campbell, H. (2013). Food and the audit society. In: A. Murcott, W. Belasco, and P. Jackson, eds., *The handbook of food research*. London: Bloomsbury, pp. 177–191.

Chang, M.C. (2014). *Expose the additives* [添加物大露出]. Available at: http://163.26.1.53/content/ActivityDetail.aspx?AID=44 [Accessed 5 Mar. 2018].

Chen, H.H. (2014). Additives and flavourings make the taste so real [食品添加物／合成香料調出「幾可亂真」的味道]. *United Daily News*, 23 Sept. Available at: https://health.udn.com/health/story/6002/365353 [Accessed 2 Mar. 2017].

Chen, H.T. (2016). Two Chefs v.s. Mangiamo. *Gradually move on with honest food*, 25 April, pp. 60–62.

Chen, S.T. (2017). Rediscover the real taste of food. *Leezen Quarterly*, 44, pp. 4–9.

Children Agro-food Education Team (2013). *The witch's recipe*. Available at: www.afa.gov.tw/cht/index.php?code=list&ids=562 [Accessed 8 Feb. 2018].

Chung, L. (2013). Concerns over food safety build in Taiwan after scandals. *South China Morning Post*, 5 Nov. Available at: www.scmp.com/news/china/article/1348274/concerns-over-food-safety-build-taiwan-after-scandals [Accessed 17 Aug. 2018].

Deleuze, G. (2006). *Foucault*. Translated from French by S. Hand. Minneapolis: University of Minnesota Press.

Draft of the Agro-Food Education Basic Law 2015 (食農教育基本法草案). Available at: https://lis.ly.gov.tw/lgcgi/lgmeetimage?cfc7cfc7cfccc8cdc5cbccd2cacf [Accessed 8 Mar. 2018].

EverydayHealth (2014). *The power of taste* [味覺力]. Available at: www.everydayhealth.com.tw/article/5381 [Accessed 22 Oct. 2016].

Food Education Teaching Plan Editing Team (2013). Food additives [食品添加物]. In: W.C. Chang, ed., *Agro-food education handbook*. Taipei: Department of Economic Development of the Taipei City Government, pp. 32–42.

Foucault, M. (1981). The order of discourse. In: R. Young, ed., *Untying the text: A poststructural anthology*. Boston: Routledge & Kegan Paul, pp. 48–78.

Foucault, M. (1988). The concern for truth. In: L.D. Kritzman, ed., *Michel Foucault: Politics, philosophy, culture. Interviews and other writings, 1977–1984*. Translated from French by A. Sheridan. New York: Routledge, pp. 255–267.

Foucault, M. (2004). *The archaeology of knowledge*. Translated from French by A. Sheridan. New York: Pantheon.

Fung, C.T. (2015). *When food safety issues happen repeatedly, let's buy directly from farmers!* Available at: www.cna.com.tw/magazine/33/201501300022-2.aspx [Accessed 25 July 2018].

Homemakers United Foundation (HUF) (2015). *The society urges the legislation of agro-food education* [民間催生食農教育立法]. Available at: www.huf.org.tw/essay/content/2901 [Accessed 22 Mar. 2017].

Homemakers United Foundation (HUF) (n.d.). *Knowing green food education* [認識綠食育]. Available at: www.huf.org.tw/action/content/1956 [Accessed 22 Mar. 2017].

Huang, J.H. (2016). Food forum appeals to eat critically and regain the dignity [品食論壇鼓勵「挑食」 籲找回吃的尊嚴]. *Liberty Times Net*, 5 Aug. Available at: http://news.ltn.com.tw/news/life/breakingnews/1786057 [Accessed 12 Oct. 2016].

Huang, L.C. (2014). *Distinguish between the real and fake soy sauces* [真醬油?假醬油?]. Available at: http://163.26.1.53/content/ActivityDetail.aspx?AID=44 [Accessed 8 Feb. 2018].

Johnston, J. and Baumann, S. (2009). *Foodies: Democracy and distinction in the gourmet foodscape*. Hoboken: Taylor & Francis.

Leezen (2016). *Knowing additives and find your own real food!* [認識添加物, 自己的「真食」自己找！] Available at: www.leezen.com.tw/article_organic.php?id=172 [Accessed 8 Sept. 2016].

Li, P.F. (2011). How to eat real food right [如何吃對真食物]. *Common Wealth Parenting*, 16, pp. 22–25.

Li, P.S. (2017). Hoh community – the good life under trees. *Gradually Move on with Honest Food* II, 27 Mar., pp. 110–115.

Lin, H.F. (2013). *Organic vegetables, you, and me* [有機蔬菜你我他]. Available at: www.afa.gov.tw/cht/index.php?code=list&ids=562 [Accessed 8 Mar. 2018].

Liu, Y.T. (2011). Incorporate the concept of TQ into nutrition and public health education. *Jen-Ai Hospital Taste Quotient Training Center*. Available at: www.tqtest.com.tw/news/index-1.asp?m=9&m1 = 9&m2 = 32&id=2 [Accessed 7 Jan. 2017].

Lu, J.Z. (2018). Taichung passes ordinances of agro-food education [台中食農教育條例過關]. *Chinatimes*, 29 May. Available at: www.chinatimes.com/newspapers/20180529000645-260102 [Accessed 10 June 2018].

Lu, P.L. (2017). Kaohsiung education bureau publishes signature products in Kaohsiung city [高市教育局出版「食在高雄樂遊遊」]. *Taiwan Times*, 3 June. Available at: www.taiwantimes.com.tw/ncon.php?num=12072page=ncon.php [Accessed 10 Jan. 2018].

Mango Social Enterprise (2016). *Pure fruit drinks* [淨果飲]. Available at: https://mango.care/juice/ [Accessed 22 Apr. 2017].

Mini Cook (2015). *Small kitchen and cool food education* [小小廚房酷食育]. Taipei: Azoth Books.

Mini Cook (n.d.). *Mini cook food education studio* [Mini cook 食育生活工作室]. Available at: https://cms.niceday.tw/minicook/ [Accessed 4 Nov. 2016].

National Chi Nan University (2017). *Local action*. Available at: www.gazette.ncnu.edu.tw/node/31 [Accessed 20 June 2018].

New Taipei City Government (2015). *New Taipei city 2015 school year food education teaching plans contest plan*. New Taipei: New Taipei City Education Bureau.

Peng, X.Z. (2014). *The only oil you can trust now is the one in your body* [現在能夠相信的「油」, 只剩下自己身上的了！]. Available at: https://health.gvm.com.tw/webonly_content_3249.html [Accessed 22 Apr. 2017].

Qiu, P. (2015). *The original taste of food: Whole Food Recipes for children and family* [原味食悟]. Taipei: My House Publishing.

Scrinis, G. (2008). On the ideology of nutritionism. *Gastronomica*, 8, pp. 39–48.

Shi, J.R. (2014). How is the taste disappeared? [味覺怎麼不見了？]. *UDN Health*, 23 Sept. Available at: https://health.udn.com/health/story/5959/366152 [Accessed 5 Feb. 2017].

Shi, Z.Y. (2014). New food labeling requirements valid from July 1st [台食品新標示　七月一日上路]. *Epochtimes*, 30 June. Available at: www.epochtimes.com/b5/14/6/30/n4189922.htm [Accessed 10 Feb. 2017].

Soyland, A.J. and Kendall, G. (1997). Abusing Foucault: Methodology, critique and subversion, history. *Philosophy and Psychology Newsletter*, 25, pp. 9–17.

Sun, W.L. (2012). Juice without fruit with misleading packaging [沒有水果的果汁涉包裝不實]. *NOWnews*, 12 July. Available at: https://m.nownews.com/news/146800 [Accessed 10 Sept. 2016].

Taiwan Food and Drug Administration (TFDA) (2010). *Proclamation of the regulations governing the labeling of packaged instant noodles* [公告訂定「包裝速食麵標示相關規定」]. Available at: www.fda.gov.tw/tc/newsContent.aspx?cid=3&id=4868 [Accessed 18 June 2017].

Taiwan Food and Drug Administration (TFDA) (2014). *Proclamation of the revision of regulations governing the labeling of packaged beverages claimed to contain fruit and/or vegetable juice* [公告修正「宣稱含果蔬汁之市售包裝飲料標示規定」]. Available at: www.fda.gov.tw/TC/newsContent.aspx?id=10843&chk=a34c63b3-360e-430e-a8d3-9cd975e1703f&param=pn%3D1%26cid%3D4%26cchk%3Df11420b2-cf8e-4d3a-beb5-66521b800453 [Accessed 18 June 2017].

Taiwan Food Safety Summit (TFSS) (2016). *Review of major food safety scandals in Taiwan*. Available at: www.twfss.org/tw/report/post/6 [Accessed 15 Feb. 2017].

Teng, S.F. (2017). Agro-food education activated, learn how to eat seriously [食農教育動起來　認真學吃飯]. *Global Views Monthly*, 19 Apr. Available at: www.gvm.com.tw/article.html?id=22854 [Accessed 18 Feb. 2018].

Tsai, Y.T and Liu, S.Y. (2017). *Signature products in Kaohsiung city, vol.3*. Kaohsiung: Education Bureau of Kaohsiung City Government.

Tu, W. and Huang, Z. (2016). Real food is the best. *Homemakers United Foundation*. Available at: www.huf.org.tw/essay/content/3472 [Accessed 9 Mar. 2017].

Ulloa, A.M. (2018). The aesthetic life of artificial flavors. *The Senses and Society*, 13(1), pp. 60–74.

Wang, C.H. (2017). What Japan taberu communications taught me. *Gradually Move on with Honest Food* II, 27 Mar., p. 16.

Wang, C.P. (2015). Wang takes care of the real tastes [王嘉平料理真滋味]. *FoodNext*. Available at: www.foodnext.net/life/lifesafe/paper/3852961741 [Accessed 15 Jan. 2017].

Wang, Y.Z. (2013). People need food safety yet government learned no lesson [人民要食安，政府卻學不會教訓]. *Global Views Monthly*, 4 Dec. Available at: www.gvm.com.tw/article.html?id=18718 [Accessed 15 Feb. 2017].

Wu, M.T. (n.d.). *New farmer model*. available at: http://modernfarmer.tainan.gov.tw/xnr_Detail.aspx?ID=abce19e2-e3b1-4fbf-a932-5deb3a0191ab [Accessed 3 Dec. 2016].

Yung, J.Y. (2018). *Food and agriculture – a mémoire of terroir to the next generation* [食農 – 給下一代的風土備忘錄]. Taipei: Guerrilla Publishing.

Zhang, Y.M. (2016). Consumption is the key to food safety [消費力　推動食安的關鍵力]. *United Daily News*, 22 Feb. Available at: http://health.udn.com/health/story/6006/1493380 [Accessed 7 Oct. 2016].

Zhu, H.F. (2013). Are you really ready for real foods? [你真的準備好吃真食物了嗎？]. *Common Health*. Available at: www.commonhealth.com.tw/article/article.action?nid=68030&fullpage=true [Accessed 1 Sept. 2016].

Chapter 9

'We need to survive'

Integrating social enterprises within community food initiatives

Andrea Tonner, Juliette Wilson, Katy Gordon and Eleanor Shaw

Introduction

This chapter documents a qualitative research study within two community food initiatives in Glasgow. This city has been highlighted as one that is particularly impoverished, with stark differences in income levels across the city. Health inequalities, over and above that which can be attributed to deprivation, exist in the city (Walsh et al., 2010). The aim of this chapter is to explore the contemporary nature of community food initiatives and their relationship with established social enterprise models. Its particular focus is on the impact of these initiatives in addressing health inequalities and how these are measured at an individual, community and socio-cultural level. The study presented forms part of a wider study focused on the impact of community social enterprises that draws on research from marketing and entrepreneurship and wider sociological literature.

This chapter begins by outlining the health inequalities derived from deprivation and discussing how Glasgow is situated in relation to these. This is followed by a discussion of the social enterprise model and community food initiatives and their use in addressing health inequalities. The remainder of the chapter reports on a study that explored understandings of social enterprise models within two community food organisations in Glasgow. It outlines the methodological choices made and key findings. It documents how two community food initiatives have developed over time, both utilising the social enterprise approach in different ways. It argues that impact measurement has not kept pace with these developments and that there needs to be a more nuanced understanding of outcomes. Conclusions drawn are that community food initiatives and social enterprises can have a key role in health improvement but only as one of a network of actors, many of whom are better placed to drive wide socio-cultural change.

Health inequalities

Health inequalities are unjust differences in health between different social or population groups. Differences in life expectancy between affluent and deprived populations are often used to illustrate these inequalities. For example, the so called 'North-South Divide' highlights the poorer health in the north of England compared to the rest of the country: a baby boy born in Manchester can expect to live for 14 fewer years in good health than a boy born in Richmond in London (Whitehead, 2014). Males living in the more affluent areas of Glasgow live 13.9 years longer than those in the more deprived areas, whilst for women the difference is 8.5 years (McCartney, 2010). Whilst comparisons between the best and worst off show dramatic differences in health, the highly influential Marmot Review (2010) also identified a consistent and systematic relationship between differences in income deprivation and disability-free life expectancy. This social gradient of health demonstrates that poor health increases with socioeconomic disadvantage. Buck and Maguire's (2015) update, using data from 2006–10, demonstrated an improvement in this social gradient, with life expectancy increasing and the difference between rich and poor lessening. However, inequalities still exist and action is needed to further reduce the gradient. Furthermore, since 2010, improvements in life expectancy have slowed: a situation that gives 'cause for alarm' (Marmot, 2017, unpaged).

Insufficient money is a highly significant cause of health inequalities (Wilkinson and Pickett, 2009), however it is not the only driver of such outcomes. For example, Glasgow experiences mortality rates beyond expectations, known as the 'The Glasgow Effect'. This is demonstrated by comparisons to Manchester and Liverpool, cities with almost identical deprivation profiles but lesser mortality; this suggests that deprivation, though a fundamental determinant of health, is only one part of a complex picture (Walsh et al., 2010). This complexity may not always be captured in sociological research on health inequalities. Scambler (2012, p. 143) notes a 'wilful denial of those "enduring social structures" like class, but not only class, that underwrite health inequalities and undermine reforms to reduce them'. This highlights a need to look further up the causal chain (Coburn, 2004).

Dahlgren and Whitehead (1991) depict the multiple potential contributory factors of health inequalities in their model of the social determinants of health (Figure 9.1). The model highlights the multi-layered nature of health determinants and depicts causal relationships between the layers. It identifies the need to explore and address what Marmot and Wilkinson (2006, p. 3) call the 'causes of the causes'. If, for example, poor health is caused by poor diet then what causes poor diet? Possible causes originate from each layer of the model: individual tastes, social norms, food availability, supply chains, cost and affordability, to name a few. The model therefore challenges the tendency to blame the sick for being sick by recognising the interplay with broader structural dynamics that fundamentally affect health.

Public health interventions that address these social determinants and therefore reduce health inequalities have become a key focus for researchers, policymakers and practitioners as they seek to evidence the efficacy of appropriate interventions. However, Bambra et al. (2010) identify two problems with the

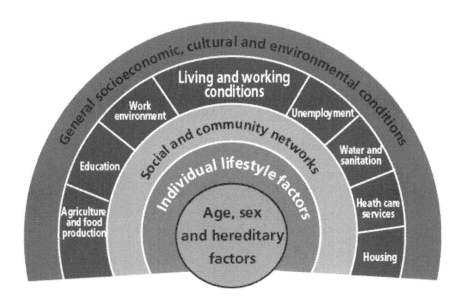

Figure 9.1 Social determinants of health model.
Source: Dahlgren and Whitehead (1991).

evidence base. Firstly, the predominance of research confirming associations between a determinant and health (e.g. job control and mental health), which can only implicitly suggest possible interventions. Secondly, research evaluating specific interventions tends to focus on modifying behaviours (e.g. smoking), minimising the determinants to individual lifestyle choices, only one layer of the Dahlgren and Whitehead (1991) model. A newer stream of research into social enterprise as a means of addressing the upstream determinants of health (e.g. Roy et al., 2012, 2014) may prove fruitful in addressing these criticisms of an individualistic focus.

At a broad level social enterprise involves the use of market-based strategies to achieve social goals (Kerlin, 2009) with key features being the primacy of social aims and the centrality of trading (Teasdale, 2011). Put simply, a social enterprise is an organisation that trades for social good. However, the reality is more complex, and although significant energy has been applied to the definition and meaning of social enterprise (Shaw and Debruin, 2013) it remains a contested concept (Teasdale, 2011). Scholarly interest in social enterprise may reflect a wider trend in sociological research which seeks to broaden the more traditional '*rational view*' that organisational behaviour is primarily driven by efficiency (Barley and Tolbert, 1997). The application of institutional theory to the field of organisations studies has drawn attention to the cultural factors that drive organisation behaviour (e.g. Scott, 2008).

Social enterprise model

Interest in social enterprise has been growing since the 1990s, becoming a focus of considerable research, policy, practitioner and educational attention (Shaw and Debruin, 2013). The social enterprise sector encompasses a diverse and heterogeneous range of organisations which adopt various structures and use multiple activities to address the needs of different client groups (Shaw and Carter, 2007). They offer an alternative type of intervention that can create a more socially embedded, equitable economy (Ridley-Duff and Bull, 2015). Social enterprise may also have the potential to challenge the neoliberal status quo (Roy and Hackett, 2017). Academic literature identifies the possible contributions of social enterprise to include making new services available, using new ways to produce traditional services, creating additional jobs, addressing specific needs of relatively small groups and communities and creation of social capital (Borzaga and Defourny, 2001). Teasdale, Lyon and Baldock (2013) note some fluidity regarding the perceived benefits of social enterprise depending on the time and context, with UK policy actors repackaging social enterprise accordingly. In 1999 they note a focus on their ability to build social capital and hence contribute to neighbourhood renewal, but more recently the focus has switched to social enterprise as innovative and entrepreneurial deliverers of public services, ultimately improving outcomes for the hardest to help. Different positioning naturally accompanies different governments. Nicholls and Teasdale (2017) note that New Labour (1997–2010) positioned social enterprise as a complementary partner to the State in welfare delivery whereas the more recent Conservative-Coalition government (2010–15) envisioned social enterprise as substitutes for State-provided public services. It was suggested this substitution would facilitate the retraction of the overbearing nanny-state, giving a greater role to communities for solving problems for themselves.

Regardless of the different positioning, the UK is seen as having the most developed institutional support structure for social enterprise in the world (Nicholls, 2010). Throughout the 2000s the social enterprise sector became the target of numerous policy initiatives that aimed to build its capacity, provide start up and growth financial aid, improve the physical asset base and provide legal recognition (Sepulveda, 2015). Although a wide and diverse range of organisations can fall under the social enterprise umbrella, Sepulveda (2015) notes a defining moment in May 2006 which moved social enterprise to the third sector. The rehoming of the Social Enterprise Unit (originally created in 2001 by New Labour to champion social enterprise across government departments) as part of the Office of the Third Sector established the role of social enterprise as a way forward for charities, most notably by reducing their dependencies on grants (Sepulveda, 2015). Accordingly, many social enterprises have emerged from pre-existing charities (Cornelius and Wallace, 2013).

However, despite what has been described in the UK as organisations 'moving towards social enterprise with the image of a tide, a force that is irresistible yet positive' (Seanor et al., 2013, p. 325), a counter narrative exists which raises concerns about the overemphasis of economic drivers (ibid), recognising the tensions between balancing market and social success. Dey and Teasdale (2013, p. 251) note how 'the seemingly unproblematic combination of these contrasting logics apparently enables social enterprises to make a profit from their activities and reinvest surpluses in the business or community'. However, as the authors suggest, the marrying of the market and philanthropic goals is not unproblematic and accordingly organisations may resist the 'tidal force'. They may have concerns that the introduction of a commercial, financially driven approach could undermine the achievement of social objectives (Doherty, Haugh, and Lyon, 2014). Alternatively, financially successful projects within a non-profit organisation may lose their competitive edge due to their social obligations. Introducing market practices into third sector organisations may undermine the traditional values of fairness and justice and compromise the organisations' ability to act as advocates or serve as conduits for free expression (Eikenberry and Kluver, 2004). The promise of social enterprise should not be taken for granted, particularly as Dey and Teasdale (2013) note that the influence of social enterprise on the third sector has been assumed rather than empirically studied; while Chaney and Wincott (2014) describe the idea that non-state organisations have superior ability to innovate as lacking empirical support.

Remaining mindful of these challenges, the role of social enterprise in addressing health inequalities warrants exploration. The 2017 Scottish Social Enterprise Census indicated that 54 percent of Scotland's 5600 social enterprises are committed to 'improving health and well-being' (Senscot, 2017, p. 5). Current literature on social enterprise and health predominantly focuses on its use in the context of health care (Roy and Hackett, 2017). This is perhaps driven by the English policy environment that sought to increase the role of the third sector in the delivery of health care particularly through the use of social enterprise 'spin-offs' from the NHS (Millar et al., 2012; Hall et al., 2012). However, Roy and Hackett (2017) take a different approach by suggesting a conceptualisation of all social enterprises having an impact on health by tackling more upstream factors, rather than by providing direct health care. Roy et al. (2014) suggest a causal pathway through which social enterprise interventions lead to the development of individual and community assets, which, in the long term, deliver improved health and well-being through increased social capital and a sense of coherence. By operating at a grass-roots level social enterprises are in an advantageous position to address the specific concerns of the local community (Cornelius and Wallace, 2013) and reciprocally embeddedness in the community should enhance credibility and garner community support (Jack and Anderson, 2002; Shaw and Carter, 2007). Social enterprises may also locate themselves in areas most at risk of health inequalities which, due to high deprivation, have few attractions for private sector activity (Bertotti et al., 2012). The most recent UK census calculated that 28 percent of social enterprises are based in the most deprived areas of the UK (Social Enterprise UK, 2017).

The empirical evidence to support these conceptualisations of social enterprises as positively impacting on health remains limited. Mason et al.'s (2015) systematic review of social innovation and health identified that studies of social enterprise were dominated by 'Work Integrated Social Enterprise' (WISE) which focuses on employability of people disadvantaged in the labour market. This concurred with Roy et al.'s (2014) earlier systematic review of social enterprise and health that identified that five of seven articles reviewed were of a WISE. The reviews concluded that social enterprise play a role in improving health primarily at individual and daily living level (Mason et al., 2015), although the evidence is limited and in need of further empirical exploration (Roy et al., 2014). Mason et al. (2015) also noted that 'social movements', in comparison to social enterprise, had more significant effects on the structural factors that drive health inequalities. Further empirical study is therefore required to consider social enterprises beyond WISE; to more robustly establish the impact of social enterprises on health; to understand the relationship between social enterprises and 'social movements' and furthering the work of Mason et al. (2015) to evidence the interplay between the layers of the social determinants model and how these may be impacted by interventions.

Community food initiatives

Community food initiatives aim to improve health, commonly through a focus on supporting a healthy diet by improving 'the five As': availability, affordability, accessibility, aptitude and attitude towards food. They also have social aims such as overcoming social isolation, promoting a sense of self-worth, empowering people, providing training, alleviating general health problems and improving the local area (McGlone et al., 1999). Aims are achieved through a wide range of activities including running food co-ops and mobile shops, cookery courses and demonstrations, food price subsidies, community cafes and community meals (Craig and Dowler, 1997). They have a range of management and organisational structures (Caraher and Dowler, 2007). A survey of 70 UK community food projects carried out in 2005 found that one-third operated as a social enterprise with a further 7 percent planning to become a social enterprise. In addition this study found that only one of those operating as social enterprise generated the majority of their income from this endeavour (Sustain, 2005). Community Food and Health (Scotland), an organisation supporting community food initiatives, produced a report in 2006 encouraging organisations to explore the potential of social enterprise recognising that, at the time, social enterprise was high on the government agenda (CFHS, 2009). The report identified benefits of social enterprise, in comparison to operating as a charity, to include providing more effective and efficient services to customers; greater independence; longer-term sustainability; stronger legal structure; greater freedom and control; easier activity planning; and greater staff security (Scottish Community Diet Project, 2006).

Despite these potential benefits, this report acknowledged a tension between balancing social and financial priorities (ibid). For example, a community café may be reluctant to remove unhealthy food from the menu due to a concern that this would deter customers. Purely focusing on nutritional health in this instance may compromise financial stability. Furthermore, different priorities for the range of stakeholders in a community food project may cause tension. For example, workers, volunteers and service users may value the social aspects and development of networks whereas commissioners consider the project's aims to be purely nutritional (Caraher and Dowler, 2007). Contrary to the 'wave of euphoria and optimism' (Bull, 2008 p. 272) surrounding general social enterprise literature, the Community Food and Health (Scotland) report acknowledged that the model may not be appropriate for everyone. It stated, 'Being more business-like and becoming enterprising is not the answer for every organisation running community food and health activities, and nor should it be' (CFHS, 2009 p. 1).

Methodology and methods

Our study adopted a case study design collecting primary data using a quasi-ethnographic approach that incorporated both interviews and observations (Gomm, et al., 2000; Pinsky, 2015; Elliott and Jankel-Elliott, 2003; Yin, 2003) with a purposive sample. Interviews were conducted with key informants in addition to ethnographic observations of everyday life and processes. In this way we endeavoured to integrate the views of different stakeholders within these community food initiatives. Interviews were also conducted with informants in key support organisations in the sector to help map the context of this research.

This study was informed by existing work on the collection of qualitative data from multiple sources (Karataş-Özkan, 2011; McAdam et al., 2014). The use of multiple methods and data to examine the views of different actors and social processes within a setting is characteristic of a case study approach. Studying individuals in context is appropriate for research aiming to move beyond the surface and explore the processes involved (Sayer, 2000). Conducting ethnographic fieldwork within community food initiatives, in addition to interviews, allowed the research team to add multiple layers of understanding during the interpretation stages (Tillman-Healy, 2003), helping to contextualise the data and clarify understanding of the experiences of informants.

The sampling criteria of the informants were that they should have experience in the management, delivery, support or use of community food projects with a strong social enterprise dimension (see Table 9.1 for informant profiles). The Scottish Index for Multiple Deprivation highlights that Glasgow is particularly impoverished in comparison to Scotland as a whole (Shipton and Whyte, 2011; SIMD, 2012). In addition, it provides evidence to show that the city presents both acute health challenges (Walsh et al., 2010) and a high saturation of social enterprises aiming to tackle the city's inequalities (GSEN and Social Value Lab, 2013). Thus, we considered Glasgow a relevant and informative site for our study.

Table 9.1 Respondent profile

	Observations	Interviews
Food Hub	Community gardeners Nutritionists Drivers Volunteers (produce sales, community cafes, food banks) End users	Manager Nutritionist Food Hub leader Volunteers
Food Supply	Food co-operatives School and nursery projects Local hospital projects Community cafes	Manager Social entrepreneur Food sales staff Volunteers
Support organisation 1		Leader
Support organisation 2		Leader

The community food sector in Glasgow encompasses a range of activities including community cooking groups, local growing projects, community cafes, food co-ops, community shops and farmers' markets. There are also network organisations offering guidance on the social enterprise approach (Sustain, 2005; Scottish Community Diet Project, 2006) and therefore we worked with two support organisations as part of our primary data collection. The first supports community food initiatives, generally advising on all aspects of management and delivery including social enterprise creation, while the second is solely focused on providing social enterprise support. We further worked with a range of informants within two community food organisations operating in areas of high multiple deprivation but with very different contextual characteristics. The identity of the participants and their respective organisations is protected by the use of pseudonyms (Hill, 1995).

Food Hub began as a food co-operative initiated by students in 2001 in response to a growth of asylum seekers being located within the area. In this area both male and female life expectancy is considerably lower than the Glasgow average. A high percentage of the population is living in income deprivation, and the proportion of children living in poverty is particularly high. Nearly a third of the population are claiming out-of-work benefits. Food Hub has 4 full-time employees, 6 part-time staff and they recruit and support 40–50 volunteers on an annual basis. The enterprise activity of Food Hub focused on the sale of fruit and veg at local *barras*[1] and only comprised a small proportion of their total income. However, at the time of the study plans were in place to undertake further business development to increase the trading side of the organisation. All other income was raised through grants from a variety of sources, usually with a focus on improving health for particular population groups.

'Food Supply' covers 10 percent of the most deprived communities in Scotland. The geographical boundary it covers is large. It includes many areas which are situated on the periphery of small towns where lack of public transport compounds difficulties for individuals in accessing amenities. The organisation has been in existence for 25 years and grew out of a federation of food co-operatives. This organisation is larger in scope, with 15 full-time employees (6 of whom are drivers), 7 part-time members of staff (many of whom are funded via community job creation schemes) and a range of volunteers. Similar to Food Hub, Food Supply's main trade was through the sale of fruit and vegetables. However, alongside small, local *barras*, Food Supply also had contracts with the local council to supply fruit and vegetables to local organisations, such as nurseries and hospitals. This wholesaling activity generated a regular and stable income source for Food Supply. Other income came from grants from a number of organisations including local councils, NHS, government and other charity partnerships.

Three researchers spent six days in the field interviewing and observing the social enterprise organisations, writing field notes and conducting more and less formal interviews. They spent a further day interviewing informants in the support organisations. Using a range of methods allowed engagement with social entrepreneurs, managers, key staff, volunteers and end users. This provided insight into the effects of social enterprise inclusion within each organisation while also considering patterns across the two organisations.

In total, seven interviews were conducted. Two interviews were within each of the community food initiatives. The first of these interviews was with the managers of each of these social enterprises and the second was with a group comprising an employee, a service user and a volunteer. The latter took the different format of a group interview as it was felt the volunteer and service users may have been more reluctant to engage without the presence of the staff member from the organisation. The interviews were non-directive whereby questions were not preplanned although the objectives of the research were known to both researchers and participants (Gray, 2004). This format was deemed appropriate as it allowed informants to talk freely around the subject (ibid). A further follow-up interview was conducted with the manager of Food Supply one year after the initial fieldwork to gain a longitudinal perspective on their organisation. Two interviews were also conducted with leaders of the two identified support organisations in order to gain a deeper understanding of the broader practices and priorities in the field.

All interviews were between 45 minutes and 1.5 hours, and with permission the interviews were audio-recorded and later transcribed. After each interview the researchers discussed initial impressions and observations, creating audio notes and research diaries to crystallise the main themes emerging (Bryman and Bell, 2007). Additional insight into the organisations and their social enterprise initiatives was gained through observations during the six days in the field and ad-hoc conversations with users and volunteers of the organisations.

The process of data analysis commenced by converting all materials into open codes in NVivo. Three team members subsequently worked together to transform the open codes into a refined set of themes and sub-themes. During this process the source data were frequently revisited to identify inconsistencies, clarify meanings and establish additional emergent codes (Fernald and Duclos, 2005).

Findings

Project origins and developments

Both community food organisations were initially set up to address a perceived unmet need, Food Hub to provide fruit and vegetables to asylum seekers in the area and Food Supply to address the gap in accessibility of food in general. Both had expanded over time and at the time of this study offered a range of additional services to meet their social objectives such as cooking groups, community gardening and emergency food aid. Although health had always been a consideration of Food Hub, it was not originally a priority for Food Supply who had been operating for a considerably longer period:

> No it historically was never healthy, never focussed on health. For instance, 15 years ago we had a pie run and we'd go round the local bakers and butchers and pick up pies and give pies to them. . . . We're now more worried about the quality of the food in those shops. So what we did ten years ago, when we started to change the name, was promote the healthy stuff rather than bar the unhealthy.
>
> (Food Supply, Manager)

The move towards a health focus had been well received by volunteers and end users, who recognised a dearth of affordable healthy food in their area and that associated cooking skills and advice can have a transformative impact on individual behaviours. Users and workers of Food Hub also valued the alternative source of health support that they felt filled a gap in their local area. When referring to the nutritional advice she had received one service user noted:

> I couldn't . . . put it this way you wouldn't get it off the NHS unless you're dropping dead and then you're dead by the time they get to you; or you couldn't do it private. I couldn't afford to do that privately.
>
> (Food Hub, Service user)

Although it was not a lack of health services that drove the initiation of the project there was a suggestion that it contributed to the ongoing requirement for some health input in that particular geographical area. By adopting a wider health focus Food Hub had broadened their appeal and relevance and had become a key broker in the community enabling access to more traditional support services.

The initial drive to meet the perceived unmet need due to poor food availability aligns with the traditional view of entrepreneurship filling a market gap. Although the origins of both organisations was to address this poor food availability their activities had expanded to address 'the five As, availability, affordability, accessibility, aptitude and attitude, the latter two now being considered "*more and more important*"' (Food Supply, Manager). This expansion had driven the requirement for a wider range of services with focus on enabling activities with population groups particularly in need, rather than just provision. One of the support organisations noted that such initiatives were constantly adapting even if the core activities appeared unchanged:

> *It's often more how they go about doing it. I mean, there must be only so many things that you can do with food . . . but there are interpretations of it that are always interesting and works with particular groups. So, it could be a youth café but a youth café with ex-offenders. . . . Sometimes the cooking classes are not new but the way they are using it is.*
>
> (Support organisation, Representative 1)

The meaning of social enterprise in the community food sector

The definition of the concept of social enterprise is contested within the literature (Teasdale, 2011). However, Food Hub had a clear interpretation of what a social enterprise is, their perception being at least one-fourth of income generated through trade, and therefore did not see themselves as fitting this self-imposed definition.

> *We don't do a lot of trading to justify being called a social enterprise. I mean, I would say you should really have twenty five percent of your income linked to trading.*
>
> (Food Hub, Manager)

This sat in contrast to the representatives from the support organisations who, considering the sector as a whole, felt that social enterprise could be much broader than many of their members recognised. They felt there was some confusion and 'fuzziness' around some of the terms and that this terminology matters little in the field:

> *We work with a lot of food co-ops and there are barely any that are actual co-operatives, in fact I can't think of any off the top of my head. But we work with community shops across the country that invariably are co-operatives but don't use the title . . . on their own branding 'community shop' is what people need to know.*
>
> (Support organisation, Representative 1)

The idea of 'what people need to know' suggests that the social enterprise element of such organisations is not a priority when creating their character or brand. Extending this, the two case studies suggested a clear distinction between their charitable and trading activities with the trading undertaken to cross subsidise their charitable operations. Food Supply operated a structure of having a separate trading arm which was set up as they felt the balance of trading verses charitable activity was, at that time, '*getting a bit out of kilter*'. In times of greater social need Food Supply felt they had to scale down their trading activity:

> *Since then the level of extreme food poverty in Scotland, in our area in particular has got so high that that equilibrium [between charity and trading] has balanced itself out again.*
>
> (*Food Supply, Manager*)

At the time of data collection, Food Hub were considering the trading opportunities available to them and were planning to recruit a business development manager to take on this role. They suggested they would need to target the '*city centre*' to generate income, therefore carrying out the trading side of the operation outside of their target geographical area for their social activities. Food Supply felt generating income from their activities, such as cooking groups, would require a change in focus, for example, classes making artisan bread. Although, they felt this might generate income it would not contribute to achieving their social goals. Trading, therefore, was considered a separate, distinct part of each organisation. Food Hub manager stated:

> *There's no way the charitable side can take a hit on any of this stuff.*
>
> (*Food Hub, Manager*)

> *[S]omeone needs to figure it all out and as far as I'm concerned it's standalone. They'll get the support from me to be a power to do it.*
>
> (*Food Hub, Manager*)

There was concern that continued development of the business arm of operations would negatively impact the social side of the organisation. Food Supply felt that for them to be entirely self-sufficient through trading they:

> *would need to put so much emphasis on developing sales that we would cease to be a charity in my view.*
>
> (*Food Supply, Manager*)

The concern that the trading activity inherent in social enterprise may negatively impact on the social objectives appears in the literature (Doherty, Haugh, and Lyon, 2014). However, contrary to the representation of social enterprises operating on a continuum between purely philanthropic goals and purely commercial goals (Dees, 1998), these case studies appeared to consider them as two entirely separate streams of activity. This conforms more closely to the representation by Roy et al. (2014) whereby the trading activity generates profits which can be invested into the social objectives. These social objectives remained the priority. The grand narrative described by Seanor et al. (2013) as organisations exhibiting a strong and positive desire to move towards being more enterprising is, similar to their findings, not representative of our organisations own perceptions. In addition, the support organisation noted that community food initiatives are often business-like even when they do not incorporate the social enterprise model:

> But actually there are lots of community food initiatives that are incredibly business like. They are not involved in major income generation but they do make sure the book balances and that's what they are bothered about. They are not looking to get any bigger, they are not looking to expand, franchise or anything like that. Whether it's a community café or anything else, they are simply looking to balance. And that doesn't balance unless you are business like. Whether you call yourself a social enterprise is secondary really.
>
> *(Support organisation, Representative 1)*

Capturing impact

Capturing the impact of the project was a crucial, if sometimes challenging task, for managers and staff. Food Supply felt their survival was dependant on evaluations, which demonstrated improvements in health:

> I don't think we'd exist without our links to the public health people at Glasgow who have come in and said 'this is working'.
>
> *(Food Supply, Manager)*

Whilst the importance of evaluating was recognised, it was sometimes a frustration for staff because it was time consuming, considered an *'admin task'*, and filling in paperwork was a distraction from the real work they wanted to be doing. Perhaps for this reason, and a perceived lack of skills, both organisations had sought help from outside agencies with evaluation work at some point. Food Supply had an established relationship with a local university who worked with them every summer to evaluate the impact of their work. This provided access to valuable and credible evaluation outputs:

> [They have] far more skills than I've got to actually do the research either because it's medical research, because it's health research or because it is statistical analysis that's required.
>
> *(Food Supply, Manager)*

Funders predominantly drove the requirement for robust evaluation data, with report-backs requiring clear evidence of how grants had been spent and what had been achieved. However, Food Hub felt the measurements taken for the purpose of completing reports did not always capture the full extent and benefit of their work:

> *There are benefits that come out of our project which we are not capturing because funders require us to report on* **their** *things.*
>
> (Food Hub, Manager)

The focus on quantitative targets for these reports, such as numbers of cookery courses or number of participants failed to capture the full impact as '*most of the work we do is very qualitative*' (Food Hub, Staff). However, it was recognised that this was changing with more opportunity to report on '*softer stuff*' (ibid). Capturing this qualitative data required a different approach than the more typical target comparison and this was undertaken through conversations and observations. Food Hub spoke of '*nuggets of gold*', from an evaluation perspective, coming from informal, natural and ad hoc conversations.

Making an impact

Given their aims, both organisations valued the difference they could make to local people's diet. Food Supply talked about the '*best*' evaluation being one that evidenced significant increases in fruit and veg consumption:

> *The best evaluation we did, the best figure, we had a 61% increase in the number of children eating fruit and vegetables during this period in the nursery.*
>
> (Food Supply, Manager)

Improvements in diet are often exemplified though increases in fruit and vegetable consumption and this increase was something of great pride to the organisation. Evidence of other changes to individual diets regularly featured in discussions, such as service users reporting eating healthier foods than previously, increased knowledge about healthy food and improved cooking skills. However, the desired targeted changes encompassed more than just the nutritional aspects of diet. Referring to one of the users of the organisations, the Food Supply manager said:

> *he said a few times in front of me you could eat food raw out the tin. I was, 'God don't say that'. Technically, he was right; it's not going to kill you but who wants to eat a tin of baked beans cold? When I heard him say that I thought 'no, we need to change that'.*
>
> (Food Supply, Manager)

However, the health benefits to individual service users stretched much further than changes in diet. Observations of a public health consultant after spending time with Food Supply concluded:

you're not a food programme at all, you've got nothing to do with diet it's public health, it's mental health, your main outcomes are more mental health.

(Food Supply, Manager)

This was based on the opportunities that the organisation offered to older people through volunteering or simply providing some company. Discussing one of the food co-ops the manager at Food Supply said:

You go in there on a Thursday and there'll always be a couple of people sitting in there. They might have shopped half an hour ago but they're still sitting talking to the volunteers or waiting on their neighbours coming in. So there's huge, huge benefits to that.

(Food Supply, Manager)

Service users were encouraged to engage with the organisation on a regular basis, facilitating the build-up of long-term support. Discussing a cookery class for older men who have never cooked whose partners now have dementia, the Food Supply manager said:

If it's just a pot of soup that's something. We're not turning that person, older person, into a cook in six weeks we're just giving the means towards the next stage of the programme.

(Food Supply, Manager)

The next stage in the programme may be further utilisation of the organisation's services or linking them in with a wider network of support such as befriending networks, community transport providers, palliative care and others. Building these networks with other services was important for the organisations and they saw food as a 'hook' for getting people involved. This unique position allowed them to access a large and wide demographic in the community. Again, this was a source of pride:

we're in schools doing growing, obviously the [fruit] barras, we're at the lunch club, we have lots of cookery going on. . . . And I was sitting doing a tally and I was like 'do you know that last week we had contact with 176 people in this area alone?' It's a lot and I was like 'wow'.

(Food Hub, Staff)

This creating and reinforcing of community was an important aspect of the work of the organisations and came through strongly in discussions:

it's actually more about linking people up.

(Food Hub, Staff)

it's the building up of community connections and working with other groups. That's definitely helping.

(Food Hub, Manager)

Service users spoke of wanting to be involved and the benefits of having opportunities to socialise:

there was this whole social thing being built up . . . and that was great, which is why Sharon said 'I want to be part of it'.

(Food Hub, Service user)

This body of evidence on the work of the organisations shows impacts at both the 'individual lifestyle' and 'social and community networks' layers of the Dahlgren and Whitehead's (1991) social determinants of health model (Figure 9.1). There is less evidence of changes to the outer two layers of the model, namely the general economic, cultural and environmental conditions. In fact, conversely, Food Supply felt one of their projects, providing emergency food aid, was necessary due to structural forces acting upon them even though they had some concerns that this activity was not an appropriate solution for people at crisis point with no food. Discussing the project, Food Supply manager said:

Our intentions were honourable. . . . I don't know if our organisation has done the right thing and are we allowing the system to be maintained? . . . But there is this whole political agenda, are you just doing what should be the government's work?

(Food Supply, Manager)

However, having run the emergency food aid programme, the manager felt they were in a stronger position to voice their concerns and engage in conversations. Had they not run the programme they would *'feel awkward'* saying that it is the wrong approach. This *'on the ground'* knowledge and understanding should allow the organisations an informed voice in discussions as to what changes are required. The representative for one of the support organisations noted how initiatives such as the case studies could be used for:

informing Government policy to say the initiatives on their own are not enough.

Community food projects, they thought were *'all part and parcel of the solution'* (Food Supply, Manager).

Discussion and conclusion

Our case studies of two community food initiatives shows that the two organisations broadened their scope over their history and became more thoughtful and wide ranging in the types of interventions they offered. These examples reflect that, in its infancy, the focus was on access to affordable food. However, reflecting shifts in governmental priorities, these initiatives had come under increasing pressure to achieve more wide-ranging outcomes. Simple food co-operatives had transformed themselves into complex organisations encompass-ing an ever-growing range of initiatives. This resulted in some very positive outcomes, including positive changes to individuals' diets, improved general health, providing opportunities to participate and socialise and strengthen-ing community ties. Our study gives empirical evidence of some of the wider social outcomes that Roy et al. (2014) suggest such initiatives can have, such as enabling and encouraging communities to be active in their own food access, making them more resilient, transforming attitudes towards healthy food and cognition of the health outcomes of diet, and developing the aptitude of both individuals and the community in preparing and handling affordable healthy food. However, this broadening of scope brings tensions within the organisa-tions as initiatives compete for internal resources and, as with any organisation in growth, the skills of those responsible are stretched to meet the increasing demands.

In the current climate, where there is considerable support for social enterprise and the positive portrayal of such models, it is perhaps surprising that there is reluctance among these initiatives to be aligned to the social enterprise vision. Our cases strongly identified themselves as being within the third sector as opposed to profit-oriented businesses, which is where these organisations situ-ate social enterprise. Despite governmental efforts to position social enterprises firmly within the third sector (Sepulveda, 2015), this has not found acceptance within these grass-roots organisations. Their strongly charitable origins and their embeddedness within charitable funding mechanisms means that their skills and networks are more closely aligned to charitable models. While they trade, the central purpose of this was to enable their charitable missions of providing low-cost healthy food as opposed to generating surplus income or having a commer-cial focus. Our interviewees acknowledged that there is considerable income to be made from the healthy food industry (e.g. selling artisan bread), but that this would only be achievable by stepping outside their target geographies or popula-tions and would be an unacceptable drift from their core social missions. They also felt that this might have detrimental impacts on their long-built reputations with their core target users and within the wider health improvement and chari-table fields. The social enterprise model may not be a good fit for every organi-sation (CFHS, 2009), and as such, it is appropriate that each initiative in any sector should find their own working business model, be that wholly charitable, wholly social enterprise or some hybrid model. Indeed, we would argue that seek-ing to define social enterprise too tightly is not particularly useful. Community food initiatives do not need to belong to a 'club' with defined rules to benefit from the principles of business with a social aim.

As noted, community food initiatives have become more ambitious in their impact over recent years seeking to address more of the 'five As' and the social

determinants of health in Dahlgren and Whitehead's (1991) model. However, it is argued here that impact measurement has not kept pace with these
changes. While these organisations saw good reasons for identifying strongly
as charities (coherent social mission and reputation) this brought demands
on resources. Reliance on charitable funding was still dominant in the cases
we studied, since their social enterprise activity remained relatively small, as
such impact had to be demonstrated on the metrics defined by their charitable
benefactors. This drove staff resources into documenting the most easily measurable and tangible individual measures such as cooking group attendance and
number of children eating fruit and vegetables. Respondents felt a frustration
that despite impacting much broader health outcomes, such as mental health
and community participation, the time and effort required to capture these
outcomes could not be as readily justified. They also faced the challenge of
designing their own metrics for such outcomes since the sector had not established clear mechanisms for capturing these kinds of impact. While they recognised that qualitative evidence was becoming more necessary in the sector,
even where not required by funders, their skills in this kind of evaluation was
lacking and much resultant reporting was ad hoc and anecdotal rather than
systematic.

Despite these measurement limitations, the broadening of community food
initiatives to encompass activities beyond their original remit has resulted in
much more impactful organisations. While there was good evidence of impact
at the individual, social and community layers of the Dahlgren and Whitehead
(1991) model, there was less evidence of impact beyond this into living conditions or socioeconomic, cultural and environmental conditions. This we suggest
is appropriate for such grass-roots initiatives, and that to expect them to have all
encompassing impact would be unreasonable. Rather they form part of a wider
ecosystem of health improvement. Their impact on the higher levels of health
derives from their experience-driven authority in the field, which they can bring
to bear on the policy-makers whom they see as having influence at these levels.
In this respect they act as a bridge between the lived experiences of health inequality in their own communities and the national initiatives required to drive
socioeconomic impact.

While these findings may reflect the experiences of other similar organisations,
their broader application cannot be assumed. This research is based on two particular community food initiatives in Glasgow who are both involved in similar
activities. Further research is needed to reveal further insights into the roles of
the impacts of social enterprise on individual and community health and well-
being and the support mechanisms needed to ensure their impact is as effective
as possible (Roy et al., 2014).

Note

1 *Barra* is a Scottish term that refers to a market-stall type outlet for selling produce. They
 are often mobile and move around different locations in a community.

References

Bambra, C., Gibson, M., Sowden, A., et al. (2010). Tackling the wider social determinants of health and health inequalities: Evidence from systematic reviews. *Journal of Epidemiology and Community Health*, 64, pp. 284–291.

Barley, S.R. and Tolbert, P.S. (1997). Institutionalization and structuration: Studying the links between action and institution. *Organization Studies*, 18(1), pp. 93–117.

Bertotti, M., Harden, A., Renton, A., et al. (2012). The contribution of a social enterprise to the building of social capital in a disadvantaged urban area of London. *Community Development Journal*, 47, pp. 168–183.

Borzaga, C. and Defourny, J. (eds.) (2001). *The emergence of social enterprise*. London: Routledge.

Bryman, A. and Bell, E. (2007). *Business research methods*. 2nd ed. Oxford: Oxford University Press.

Buck, D. and Maguire, D. (2015). *Inequalities in life expectancy: Changes over time and implications for policy*. London: Kings Fund.

Bull, M. (2008). Challenging tensions; critical, theoretical and empirical perspectives on social enterprise. *International Journal of Entrepreneurial Behaviour and Research*, 14(5), pp. 268–275.

Caraher, M. and Dowler, E. (2007). Food projects in London: Lessons for policy and practice – a hidden sector and the need for 'more unhealthy puddings . . . sometimes'. *Health Education Journal*, 66, pp. 188–205.

Chaney, P. and Wincott, D. (2014). Envisioning the third sector's welfare role: Critical discourse analysis of 'post-devolution' public policy in the UK 1998–2012. *Social Policy & Administration*, 48, pp. 757–781.

Coburn, D. (2004). Beyond the income equality hypothesis: Class, neoliberalism, and health inequalities. *Social Science & Medicine*, 58(1), pp. 41–56.

Community Food and Health Scotland (2009). *Minding their business too*. Available at: www.communityfoodandhealth.org.uk/wp-content/uploads/2009/09/cfhsmindingtheir ownbusinesstoo-3421.pdf/ [Accessed 21 Mar. 2018].

Cornelius, N. and Wallace, J. (2013). Capabilities, urban unrest and social enterprise: Limits of the actions of third sector organisations. *International Journal of Public Sector Management*, 26, pp. 232–249.

Craig, G. and Dowler, E. (1997). Let them eat cake! Poverty, hunger and the UK State. In: G. Riches, ed., *First world hunger food security and welfare politics*. Basingstoke: Macmillan Press Ltd.

Dahlgren, G. and Whitehead, M. (1991). *Policies and strategies to promote social equity in health*. Stockholm: Institute for Future Studies.

Dees, J.G. (1998). Enterprising non-profits. *Harvard Business Review*, 76(1), pp. 55–67.

Dey, P. and Teasdale, S. (2013). Social enterprise and dis/identification. *Administrative Theory & Praxis*, 35, pp. 248–270.

Doherty, B., Haugh, H. and Lyon, F. (2014). Social enterprises as hybrid organizations: A review and research agenda. *International Journal of Management Reviews*, 16, pp. 417–436.

Eikenberry, A. and Kluver, J. (2004). The marketization of the nonprofit sector: Civil society at risk? *Public Administration Review*, 64, pp. 132–140.

Elliott, R. and Jankel-Elliott, N. (2003). Using ethnography in strategic consumer research. *Qualitative Market Research: An International Journal*, 6(4), pp. 215–223.

Fernald, D.H. and Duclos, C.W. (2005). Enhance your team-based qualitative research. *The Annals of Family Medicine*, 3(4), pp. 360–364.

184 Andrea Tonner et al.

Glasgow Social Enterprise Network and Social Value Lab (2013). *Social enterprises in Glasgow: Scale as well as substance*. Available at: www.gsen.org.uk/ [Accessed 21 December 2017].

Gomm, R., Hammersley, M. and Foster, P. (eds.) (2000). *Case study method*. London: Sage.

Gray, D.E. (2004). *Doing research in the real world*. London: Sage Publications.

Hall, K., Miller, R. and Millar, R. (2012). Jumped or pushed: What motivates NHS staff to set up a social enterprise? *Social Enterprise Journal*, 8, pp. 49–62.

Hill, R.P. (1995). Researching sensitive topics in marketing: The special case of vulnerable populations. *Journal of Public Policy & Marketing*, 14, pp. 143–148.

Jack, S. and Anderson, A. (2002). The effects of embeddedness on the entrepreneurial process. *Journal of Business Venturing*, 17, pp. 467–487.

Karataş-Özkan, M. (2011). Understanding relational qualities of entrepreneurial learning: Towards a multi-layered approach. *Entrepreneurship & Regional Development*, 23(9–10), pp. 877–906.

Kerlin, J. (2009). *Social enterprise: A global comparison*. New England: University Press of New England.

Marmot, M. (2010). *Fair society, healthy lives: The Marmot review*. Available at: www.gov.uk/dfid-research-outputs/fair-society-healthy-lives-the-marmot-review-strategic-review-of-health-inequalities-in-england-post-2010. [Accessed 7 Mar. 2018].

Marmot, M. (2017). *The rise of life expectancy in the UK is slowing*. Available at: http://marmot-review.blogspot.co.uk/2017/07/the-rise-of-life-expectancy-in-uk-is.html. [Accessed 8 Mar. 2018].

Marmot, M. and Wilkinson, R. (eds.) (2006). *Social determinants of health*. 2nd ed. Oxford: Oxford University Press.

Mason, C., Barraket, J., Friel, S., et al. (2015). Social innovation for the promotion of health equity. *Health Promotion International*, 30, pp. ii116–ii125.

McAdam, R., Antony, J., Kumar, M., et al. (2014). Absorbing new knowledge in small and medium-sized enterprises: A multiple case analysis of six sigma. *International Small Business Journal*, 32(1), pp. 81–109.

McCartney, G. (2010). Illustrating health inequalities in Glasgow. *Journal of Epidemiology & Community Health*, 65(1), p. 94.

McGlone, P., Dobson, B., Dowler, E., et al. (1999). *Food projects and how they work*. Joseph Rowntree Foundation. New York: New York Publishing Services.

Millar, R., Hall, K. and Miller, R. (2012). Right to request social enterprises: A welcome addition to third sector delivery of health care? *Voluntary Sector Review*, pp. 275–285.

Nicholls, A. (2010). Institutionalizing social entrepreneurship in regulatory space: Reporting and disclosure by community interest companies. *Accounting, Organizations and Society*, 35, pp. 394–415.

Nicholls, A. and Teasdale, S. (2017). Neoliberalism by stealth? Exploring continuity and change within the UK social enterprise policy paradigm. *Policy & Politics*, 45, pp. 323–341.

Pinsky, D. (2015). The sustained snapshot: Incidental ethnographic encounters in qualitative interview studies. *Qualitative Research*, 15(3), pp. 281–295.

Ridley-Duff, R. and Bull, M. (eds.) (2015). *Understanding social enterprise: Theory and practice*. 2nd ed. London: Sage Publications.

Roy, M., Donaldson, C., Baker, R., et al. (2012). *Social enterprise: New pathways to health and well-being*. Journal of Public Health Policy, 34(1), pp. 55–68.

Roy, M., Donaldson, C., Baker, R., et al. (2014). The potential of social enterprise to enhance health and well-being: A model and systematic review. *Social Science & Medicine*, 123, pp. 189–193.

Roy, M. and Hackett, M. (2017). Polanyi's 'substantive approach' to the economy in action? Conceptualising social enterprise as a public health 'intervention'. *Review of Social Economy*, 75, pp. 89–111.

Sayer, A. (2000). *Realism and social science*. Thousand Oaks, CA: Sage.

Scambler, G. (2012). Health inequalities. *Sociology of Health & Illness*, 34(1), pp. 130–146.

Scott, W.R. (2008). *Institutions and organizations: Ideas and interests*. 3rd ed. London: Sage.

Scottish Community Diet Project (2006). *Minding their own business*. Glasgow: Scottish Consumer Council.

Scottish Index of Multiple Deprivation (2012). *Local authority summary – Glasgow city*. Available at: www.gov.scot/Topics/Statistics/SIMD/Publications/LASummariesSIMD 12/LASummaryGlasgowCity12 [Accessed 27 July 2017].

Seanor, P., Bull, M., Baines, S., et al. (2013). Narratives of transition from social to enterprise: You can't get there from here! *International Journal of Entrepreneurial Behaviour and Research*, 19(3), pp. 324–343.

SENSCOT (2017). *Diet, activity and healthy weight: The role of social enterprise*. Available at: https://senscot.net/wp-content/uploads/2017/11/Senscot-Briefing-Diet-Activity-and-Healthy-Weight.pdf. [Accessed 8 Mar. 2018].

Sepulveda, L. (2015). Social enterprise – a new phenomenon in the field of economic and social welfare? *Social Policy & Administration*, 49, pp. 842–861.

Shaw, E. and Carter, S. (2007). Social entrepreneurship: Theoretical antecedents and empirical analysis of entrepreneurial processes and outcomes. *Journal of Small Business and Enterprise Development*, 14, pp. 418–434.

Shaw, E. and De Bruin, A. (2013). Reconsidering capitalism: The promise of social innovation and social entrepreneurship? *International Small Business Journal*, 31, pp. 737–746.

Shipton, D. and Whyte, B. (2011), *Mental health in focus: A profile of mental health and wellbeing in greater Glasgow & Clyde*. Glasgow: Glasgow Centre for Population Health.

Social Enterprise UK. (2017). *The future of business: State of social enterprise survey*. Available at: www.socialenterprise.org.uk/Pages/Category/state-of-social-enterprise-reports [Accessed 13 Mar. 2018].

Sustain (2005). *Social enterprise for community food projects: A solution to the funding conundrum or just another fashionable magic bullet? A policy briefing paper*. Available at: www.sustainweb.org/pdf/PolicyBriefng_05.pdf [Accessed 6 June 2017].

Teasdale, S. (2011). What's in a name? Making sense of social enterprise discourses. *Public Policy and Administration*, 27, pp. 99–119.

Teasdale, S., Lyon, F. and Baldock, R. (2013). Playing with numbers: A methodological critique of the social enterprise growth myth. *Journal of Social Entrepreneurship*, 4, pp. 113–131.

Tillman-Healy, L.M. (2003). Friendship as method. *Qualitative Inquiry*, 9, pp. 729–749.

Walsh, D., Bendel, N., Jones, R., et al. (2010). It's not just deprivation: Why do equally deprived UK cities experience different health outcomes? *Public Health*, 124, pp. 487–449.

Whitehead, M. (2014). *Due north: The report of the inquiry on health equity for the north*. Manchester: Centre for Local Economic Studies. Available at: www.cles.org.uk/publications/duenorth-report-of-the-inquiry-on-health-equity-for-the-north/ [Accessed 7 Mar. 2018].

Wilkinson, R. and Pickett, K. (2009). *The spirit level: Why more equal societies almost always do better*. London: Allen Lane.

Yin, R. (2003). *Applications of case study research*. London: Sage.

Index

Printed and bound by CPI Group (UK) Ltd, Croydon, CR0 4YY

24/10/2024

01778281-0002